T0340127

GRAPHITIC NANOFIBERS

GRAPHITIC NANOFIBERS

A Review of Practical and Potential Applications

JUZER JANGBARWALA

CEO, Voltek Energy, Inc., Santa Margarita, CA, United States

ELSEVIER

AMSTERDAM • BOSTON • HEIDELBERG • LONDON • NEW YORK • OXFORD
PARIS • SAN DIEGO • SAN FRANCISCO • SINGAPORE • SYDNEY • TOKYO

Elsevier
Radarweg 29, PO Box 211, 1000 AE Amsterdam, Netherlands
The Boulevard, Langford Lane, Kidlington, Oxford OX5 1GB, United Kingdom
50 Hampshire Street, 5th Floor, Cambridge, MA 02139, United States

Notices

Knowledge and best practice in this field are constantly changing. As new research and experience broaden our understanding, changes in research methods, professional practices, or medical treatment may become necessary.

Practitioners and researchers must always rely on their own experience and knowledge in evaluating and using any information, methods, compounds, or experiments described herein. In using such information or methods they should be mindful of their own safety and the safety of others, including parties for whom they have a professional responsibility.

To the fullest extent of the law, neither the Publisher nor the authors, contributors, or editors, assume any liability for any injury and/or damage to persons or property as a matter of products liability, negligence or otherwise, or from any use or operation of any methods, products, instructions, or ideas contained in the material herein.

British Library Cataloguing-in-Publication Data
A catalogue record for this book is available from the British Library

Library of Congress Cataloging-in-Publication Data
A catalog record for this book is available from the Library of Congress

ISBN: 978-0-323-51104-9

Publisher: Matthew Deans
Acquisition Editor: Simon Holt
Editorial Project Manager: Sabrina Webber
Production Project Manager: Kiruthika Govindaraju
Cover Designer: Greg Harris

Typeset by MPS Limited, Chennai, India

To my wife Farzana
The center of my universe.

To my sons Mustafa & Ali
An apple for each eye.

To my (late) father Moiz
"Confidence is the first step towards success"...
Made me an entrepreneur.

To my mother Vaza
"Education first! Knowledge guides the soul"...
Made me a student for life.

To my (late) father-in-law Shafi-ul-Huq
"I believe in you" ......
I stood up after many a fall.

To my mother-in-law Shamim
"You didn't have to be born of my womb to be my son"....
What can I say?

To my siblings –
They say one doesn't have a choice..........
I couldn't have chosen any better.

To my daughter-in-law Dana
Finally! We have a daughter!

To my family & friends
Thank you for putting up with my eccentricities.

To my country America
Not perfect, but by far the best.

CONTENTS

LIST OF FIGURES

LIST OF TABLES

BIOGRAPHY

Juzer Jangbarwala is currently the CEO of Voltek. His work spans over 35 years, primarily in commercializing innovative technologies. He has established water treatment companies around his own patents and has always been intimately involved with the technical as well as the business aspects of his companies. To commercialize his inventions, he started Catalyx, a technology incubator in 2001, which is when he got involved with graphitic nanofibers. Many of the technologies developed involved graphitic nanofibers either directly or indirectly. He established Voltek in 2015. Among the IP portfolio of Voltek are 5 issued patents and several pending patents utilizing graphitic nanofibers in the fields of water and wastewater treatment, waste to fuels and catalysis. He has a BS in Chemical Engineering from Lehigh University, Bethlehem, PA and holds more than 20 patents.

Juzer Jangbarwala
Voltek Energy, Inc., Santa Margarita, CA, United States

FOREWORD

I have had the pleasure and honor to work with Juzer for 3 years but have known him as a friend, mentor, and my own personal "Wikipedia" for 15 years. Juzer is one of those technically gifted individuals that comes along infrequently, with the capacity to disrupt fundamental technical concepts and firmly held scientific dogma. Distinguishing him from the typical innovator and the typical super-specialist is his ability to see the practical implications, the utility, and the commercialized application of a technology or invention. If I was given the challenge to conquer the Universe, Juzer will be my chief technologist, chief engineer, and chief innovator.

Cornelius J. Kriek
CEO BOC Edwards Materials & Services (Ret)

Are we there yet? ....Are we there yet?

Like parents, the prolific sensational announcements about prodigious paradigm shifts by graphene and nanotubes obfuscate reality by implying "we are almost there." The barrage of information and the engraphic effect forces cognition to take a simpler, faster path and irrational expectations generated by the brain's *System 1* to act without detailed analysis typically carried out by *System 2* [1]. No doubt these materials can potentially provide immeasurable benefits for almost all walks of life. But do we have the means to produce these materials within the practical constraints of manufacturing, economics, safety and market acceptance, all in the absence of any benchmarks? Could we achieve these benefits with *less* glamorous and *less* headline grabbing materials?

My vexation at the incertitude created by the hype and the implied unrealistic timelines finally motivated me to pen down my thoughts, knowledge, and experience. In Nate Silver's words, I wanted to attempt to separate the "signal" from the "noise" [2] to the best of my ability, and by doing so, hopefully *give encouragement to, and inspire creative thought in the next generation of engineers and product development specialists.*

Though the subject knowledge does not lend itself to be disseminated in simple language, I have endeavored to present a holistic view in a coherent and easy-to-understand manner getting technical only when necessary. With the underlying assertion that graphitic nanofibers can be manufactured economically, I draw parallels between graphitic nanofibers and published works for "graphene," "graphene oxide," and a myriad of materials with graphene-bearing names. The field of carbonaceous nanomaterials is vast, and I do not profess to know all the materials in enough depth to include them in this book. I have, in the past, used GNF as an acronym, but have decided to switch to GN to avoid confusion with yet another popular name, graphite nanoflakes (GNF).

To some who know me, I may well be identified as a cynic, certainly with many crotchets. For those readers in the academic circles, some of my innuendos may be misconstrued as an indictment of the field of research and development itself. It is not my intent to belittle fundamental research. Rather, it is to share, without prejudice, this important topic from a different vantage point. It behooves us all to point out the instances when the unspoken, sacred line is crossed by trusted professions for commercial gains [3,4], without regard to accuracy [4,5] especially if there are deleterious effects from such actions [3]. I plan to enjoy the journey and make unabated comments about my view of causes and effects. So I apologize in advance for my unconventional style.

INTRODUCTION

This book is meant to be an easy read. I wanted to give the reader a perspective from a different vantage point; a holistic view that encompasses sufficient information on the science, engineering, and economics of an often misunderstood field. It is light in all three disciplines, but hopefully descriptive enough to get the point across, without extensive discussions of any one topic. I like thin books that get to the point, yet are not very esoteric, so I have endeavored to keep this one short and simple.

If the reader desires detailed derivations, mathematical theories, and models, this is not the book, though some are discussed in great detail, when necessary. Neither is it for someone who desires detailed engineering design or Wall Street style financial analyses. It is a simple book from an author with a simple mind, and an average IQ. The ultimate goal is to show how some basic, yet remarkable materials have been ignored because of their lack of complexity and hype potential. In some instances, where not relevant, I expect the reader to find other sources for more detailed historical aspects of discovery, first applications, etc., which is generally covered by most technical articles and books, all too redundantly, in my opinion. Instead, I shall give very brief backgrounds and jump straight into a discussion of the characteristics at hand.

Having said that, the nature of the subject still demands a grasp on the basic physical chemistry concepts if the reader wants to understand my deductions for substitution of graphitic nanofibers (GN) for the more exotic materials. Those who are looking at the gist of things will also be able to achieve their objective by reading the summary following most chapters.

In Chapter 1, Brief Overview of Carbon and Its Cousins, some relevant chemistry and physics is reviewed.

The substitution argument to use GN in place of other exotic materials requires a some understanding of the special role of the electronic structure of elemental carbon, and the presence of reactive sites, and the special role these qualities play at the nano level and make things possible that macro materials cannot.

In Chapter 2, Review of Carbonaceous Nanomaterials and Graphite, having laid out the basics of atomic structure, I will expand on the characteristics of GN as well as those of the other carbonaceous materials starting with graphite, then carbon nanotubes (CNTs), GO, and Graphene.

This chapter will therefore give specific insight into these characteristics and set the stage again, for the comparison between the materials and then the substitution argument to follow.

From Chapter 3, Graphitic Nanofibers—The Path to Manufacturing, I focus on different aspects of GN, presenting the process from start to finish. I will start with various catalysts, followed by a look at the reaction mechanisms driving the formation of nanomaterials. This discussion will give us some insight into factors that may affect design of reactors. I end that chapter with an interesting reference that discusses growth rates to give food for thought for designers.

Chapter 4, Manufacturing, will delve into the synthesis of GN.

Different types of commercially operating reactors and designs will be discussed in relevant detail. Nanomaterials have raised questions about health effects. Since health concerns are mostly based on manufacturing environments, health effects about nanomaterials, concerns, and status are also covered in this chapter.

In Chapter 5, Costs of Manufacturing, I present a detailed look at the cost of manufacturing CNTs, GO, and GN and derive a comparison table. The cost derivation follows a real-world business approach to determine costs incurred by manufacturers. With the costs established for each material, economic viability will be determined for some applications promoted currently, sometimes with just a back of the napkin calculation.

Chapter 6, Functionalization and In Situ Polymerization, is a prelude to our discussion of applications, with some thought given to the functionalization of GN, and the possibility that GN being viable in many types of in situ polymerization reactions, where the other famous materials do not appear realistic for use by industry, especially due to their costs.

Chapter 7, Applications, will be dedicated to a few applications that I feel practical to consider today, if GN were to be substituted for graphene, graphene oxide, and other materials.

Lithium–ion batteries, catalysis, water and wastewater treatment, gas purification, heat transfer, lubrication and drilling fluids are discussed with this objective in mind.

Finally, I will attempt to summarize the material covered in the book, and some thoughts and opinions.

I hope this book provides some guidance to the curious minds with a simple but comprehensive view of the path to commercialization and what can be possible in the near future. I reiterate. It is important for the reader to understand the intent of this book, which is to provide tools for and to encourage novelty, cross-pollination, and original thought to further the use of nanomaterials in industry. It is not a technology recipe book, neither is it an attempt to show academic prowess. I offer my thoughts for the possibilities of the use of GN in the applications discussed. The source of material for the subject matter in this book varies:

1. Published technical literature specifically covering applications of GN.
2. Published technical literature for applications using CN, CNTs, graphene, and GO.
3. Lastly, where appropriate, I share my personal experience with these extraordinarily versatile and useful materials.

CHAPTER ONE

Brief Overview of Carbon and Its Cousins

1.1 ELECTRONIC STRUCTURE

Carbon is an exceptional element with remarkable properties, but most people forget the versatility of this key component of life itself. Reviewing these basic properties will help you appreciate our discussions later. Those of you who remember their high school chemistry may skip this section. Essential to all chemical reactions is the bonding mechanism and electronic structure of the reactants. Normally, chemical bonding takes place between similar structures. Two s-orbitals or two similar p-orbitals bind together in an antibonding and bonding manner. For many of the carbon atoms, the binding of the same kind of molecular orbitals is not the case. In fact they form bonds by mixing (hybridizing) different orbitals, namely s- and p-orbitals. This versatility of carbon–carbon bonding creates a wealth of extraordinary physical properties.

A carbon atom contains six electrons which occupy the following electron configuration: $(1s)^2(2s)^2(2p)^2$. In the ground state there are two unpaired electrons in the outer shell, seemingly possessing the ability to bind only two additional electrons, but, as we very well know, a four electron binding behavior is observed. The reason is that one electron from the 2s- state can easily be excited to the 2p- state due to the very small energy difference; this can result in a mixed state formed out of one s- orbital and three p-orbitals, namely p_x, p_y, and p_z in which case four new hybrid orbitals can be formed and termed sp^3 orbitals (see Fig. 1.1). This combination of the hybrid orbitals produces a tetrahedral structure we know as a diamond. Due to the three-dimensional sp^3 structure, the binding strength between neighboring carbon atoms is equal for each atom and is very strong.

When one s-orbital and only two p-orbitals *hybridize*, they form a planar assembly with a characteristic angle of 120° between the hybrid orbitals. The hybridized orbitals form σ-bonds in the planar direction.

Graphitic Nanofibers.
DOI: http://dx.doi.org/10.1016/B978-0-323-51104-9.00001-7

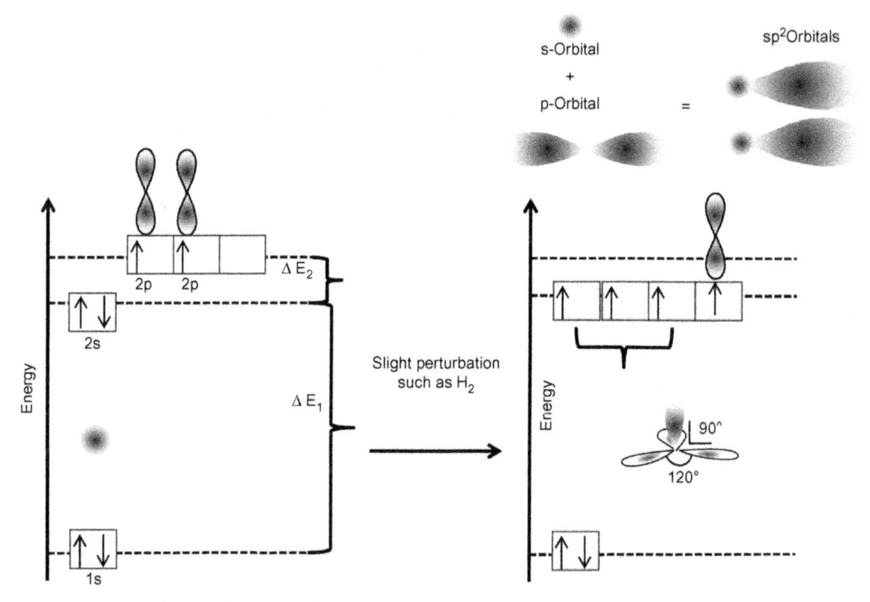

Figure 1.1 Carbon orbitals, hybridization.

The additional p-orbital exists perpendicular to the sp^2-hybrid orbitals and forms π-bonds. In graphite, this planar structure is formed between 6 carbon atoms and the p-orbitals form π-bonds which create a delocalized state for all the electrons. Such a delocalized state in graphite can be thought of as a donut-shaped electron density above the planar surface.

The in-plane σ-bonds in sp^2-hybridized carbon are stiff against longitudinal forces but soft for angular deformations. This flexibility opens the door for the wide range of stable graphitic nanostructures. One major property is in electron transfer. The electronic properties of sp^2-hybridized carbon materials are generally determined by the delocalized π-orbital. Unlike the strongly bonding σ-orbitals, the π-orbitals are generally close to the Fermi energy. The Fermi energy is the difference between the highest and the lowest occupied orbitals of isolated particles at absolute zero temperature (hence a measurement of the "total" chemical potential of the electron), leading to the in-plane conducting or semiconducting properties of graphitic materials. Many electronic applications can benefit from this property. To study the orbitals, we need to imagine a single sheet of planar carbon material that is two dimensional in nature, and (now famously) known as graphene. In a single, flat graphene sheet, the in-plane sp^2 orbitals are symmetrical and the π-orbitals are asymmetrical and delocalized. This discourages any matrix elements that can couple both types of orbitals. These two groups of orbitals form two independent sets of bands: (1) the σ-bands, formed by the sp^2-hybridized orbitals, which are responsible for the structure. These lie far below and above the Fermi energy, so they have negligible influence on the electronic properties at energies relevant for electronic transport. (2) The π-bands, formed by the atomic π-orbitals, are half filled, right around the Fermi energy, and are responsible for transport and other electronic properties at low energy.

Therefore, on a finite dimension, at the end of this plane, the edges of the perfect hexagon structure end up with incomplete σ-bonds, commonly referred to as "dangling bonds." These dangling bonds impart important qualities to the graphene plane, which are very relevant to the applications we are about to explore. I have therefore dedicated a separate section to explain these states in detail next.

To summarize, graphite is a multi layered planar structure. Each individual layer is called graphene and is formed by three hybridized sp^2 orbitals. These orbitals form σ-bonds, in the planar direction, creating a sheet that is very stable longitudinally. The p-orbital electron, which is not hybridized exists in a direction perpendicular to the plane and is delocalized with the other p-orbital electrons of the five neighbor carbon atoms. This property is responsible for a very important quality, desired by electronic transport applications. A singular sheet in graphite is called graphene and consists of carbon atoms attached by σ-bonds and delocalized p-orbitals. When an infinite graphene sheet is theoretically cut, the edges from the planar σ-bonds are left with dangling bonds.

1.2 THE EDGE SITES

The importance of an edge to a graphene sheet parallels that of a surface to a crystal. As is common knowledge, the lateral dimension and plane of the graphene sheet is being pursued for future electronic and thermal transfer applications. Obtaining such sheets in practical dimensions without any surface defects is providing scientists with unprecedented opportunities in physics and chemistry research. But pristine graphene sheets with dimensions even in the single digit μM size, have proven to be far more difficult to synthesize than initially assumed. Outside of prospective applications that promise to enhance electrical conductivity, semiconductor behavior, and thermal conductivity (which, currently are stuck in the (inability to produce in larger) size conundrum indefinitely, very few applications are candidates for this expensive material. Whether you want to make a plastic composite, a higher performance catalyst, a better lubricant, a better drilling fluid, etc., the most important requirement turns out to be, not single or few layer "graphene", but the ability of the nanomaterial flake(s) to either react or be miscible in the material to be enhanced. Not surprisingly, then, the graphene applications being advertised, as near term are primarily those that would result from the unique properties of the edge sites, not those of a single (or few)-layer graphene. No experimental data within these large scale applications to my knowledge suggests that specifically single-layer or few-layer [6–10] graphene materials are required to achieve the results reported by researchers using expensive graphene. Even if there was such a claim, my contention is that graphitic nanofibers (GN) will be able to achieve a high percentage of those enhancements, resulting in better economic returns. Since pristine graphene sheets of usable size are not within *my* definition of "practical" (yet), I will focus on the many applications made possible by these edge states. Therefore a detailed discussion about energy levels, bonding abilities, and bond strengths of these edge sites is crucial in my endeavor to demonstrate the versatility of GN.

So let us imagine cutting through an infinite graphene sheet. We first end up first breaking the (C–C) σ-bonds and then obtaining two semi-infinite graphene sheets, each with a one-dimensional edge. Along the cut, the edges will be exposed. The cut also introduces a boundary at the edge to the previously fully delocalized π-electron system. Imagine a

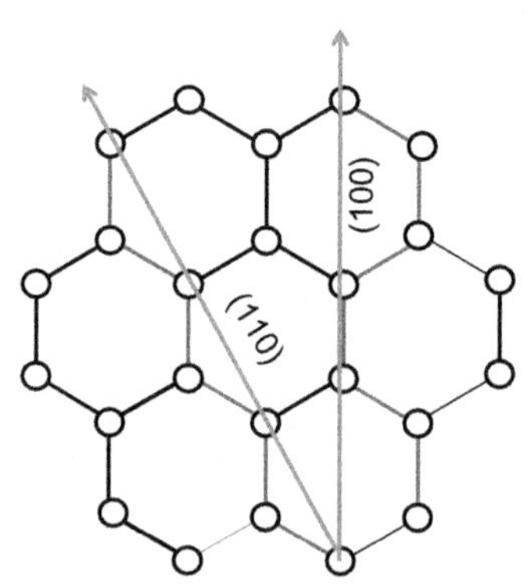

Figure 1.2 ⟨100⟩ ⟨110⟩ direction cuts across graphene plane to form zigzag and armchair faces. See also Fig. 1.3. The exposed electrons and sites are in red (dark grey in print versions).

hexagonal plate, with a donut suspended above. If cut, half the plate as well as half of the donut will disappear. The edges of the plate will have no carbons to bond with, and similarly, neither will the donut.

I will give a brief and simplified view of quantum mechanics to explain a reference following. The geometry of the edge sites makes a major difference in the π-electron structure at the edge. In quantum mechanics, an electron mass can be represented as (short bursts of) waves, and the complex formula for the resulting plot ("shape") is called the wave function. The location of the electron via probabilistic calculations can be estimated based on this wave function. At equilibrium, the band will be represented by a flat line representing the Fermi level. The energies of the bands are calculated in what is called "k-space" or sometimes called "momentum space". This is an abstract space intimately related to the real or position space hence a k vector is then conveniently used to calculate the energies of the extended orbitals in crystalline solids. Again, this definition is only to give you some background of the techniques used in the next reference to verify the energy difference between the two exposed faces. A detailed discussion is out of the scope of this book. Nakada et al. [11] showed that the zigzag edge in a graphene sheet has a flat band near the Fermi level for the k vector between $2\pi/3$ and π. For $k = \pi$, the wavefunction becomes completely localized at the edge sites. This flat-band feature and its corresponding localized state are unique to the zigzag edge (they are completely absent from the armchair edge).

Edge or surface energies both quantify the disruption of interatomic bonds. If all dangling bonds were equal in graphene, the edge energy proportional to their density would be higher for the more tightly packed armchair than for the less dense zigzag, by exactly a factor of $2/\sqrt{3} = 1.15$. In Fig. 1.4 [12], however, this very difference in density

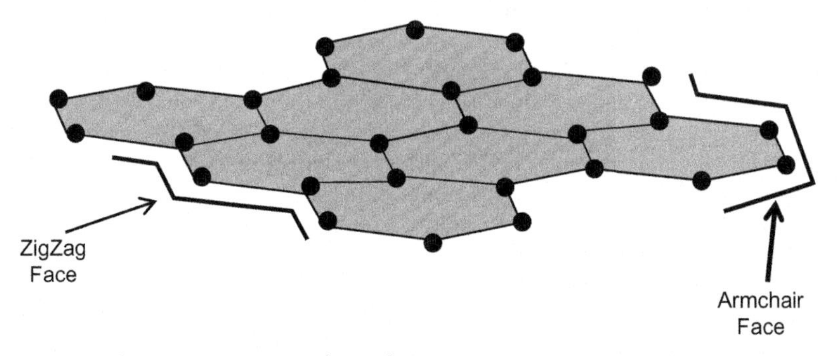

Figure 1.3 Edges sites zigzag and armchair.

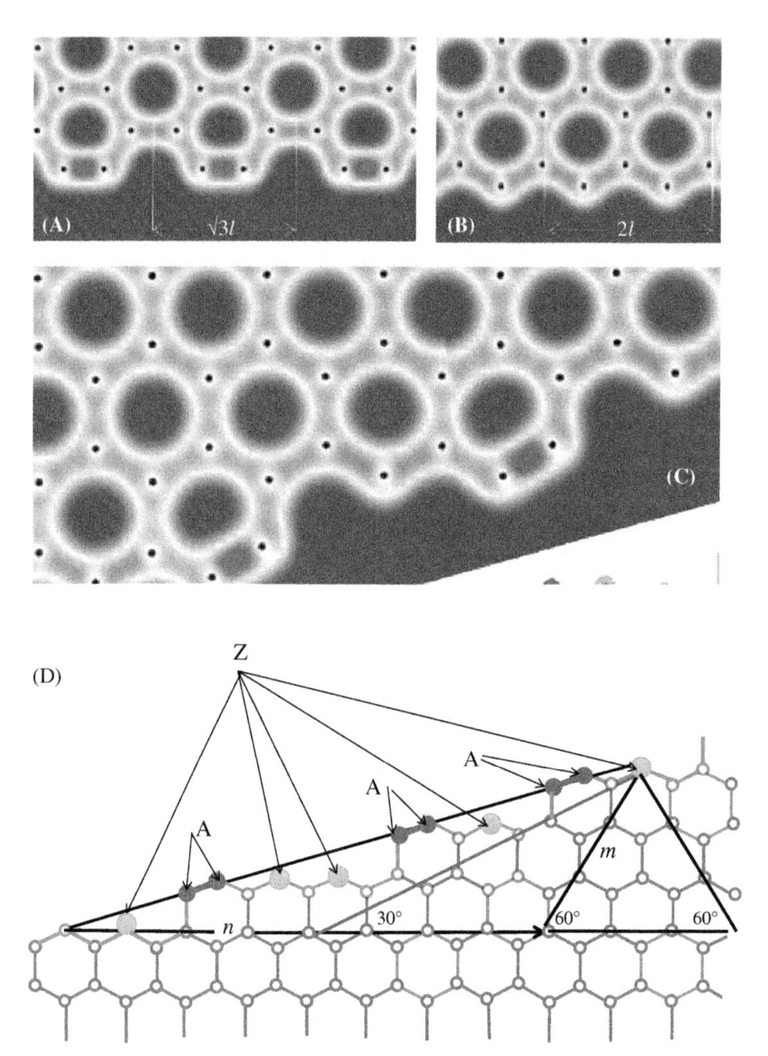

Figure 1.4 Different atomic spacing along the armchair (A) and zigzag (B) edges results in distinctly different electron density distributions, with armchair edge atoms forming shorter and stronger triple bonds. This distinction between the two types of atoms is preserved in a mixed chiral edge (C), as the computed electron density illustrates (from blue (dark gray in print versions) for zero up to red (light gray in print versions) for the highest value). Schematics of the edge (D) along the $(n-m)$ direction assists the atom counting: 2m A-atoms (count along the red (dark gray in print versions) line at 30°), and (n-m) Z-atoms (count along the horizontal black line segment). Dividing these numbers by the length $(n^2 + nm + m^2)^{1/2}$ of the edge (the diagonal on the left) yields the necessary densities, c_A and c_z in this example of the (8,3) edge. So there are 6 A and 5 Z atoms. *From Y. Liu, A. Dobrinsky, B.I. Yakobson, Graphene edge from A to Z—and the origins of nanotube chirality, Phys. Rev. Lett. 105 (2010), 235502. Please note: Panel (D) has been modified to explain the intended concept with labeling.*

makes armchair sites form triple bonds and thus lowering their energy relative to the zigzag edges.

Fujii and Enoki [13] also write that armchair edges are energetically stable due to the aromatic stability, whereas less-aromatic zigzag edges are unstable. The nonbonding edge state of π-electrons is confirmed to exist in the zigzag edges, in spite of the absence of such a state in armchair edges. In a finite-length zigzag edge embedded between armchair edges, the electron confinement effect is observed in the edge state.

Kobayashi [14] predicted the existence of such a flat band and localized state on a zigzag-edged vicinal graphite surface. Highly oriented pyrolytic graphite (HOPG) surfaces have also shown the existence of the zigzag edge, and the localized state at the zigzag edge was confirmed by scanning tunneling spectroscopy (STS), which essentially takes images at the atomic level.

Now, if we imagine cutting the semi-infinite graphene sheet again, and in the same direction as the first cut, parallel to the edge, we would generate what is now known as a graphene ribbon with two zigzag edges. If the ribbon width falls within the nanometer range (zigzag graphene nanoribbons, GNR), the two edges would interact with each other, possibly forming a single-walled carbon nanotube.

Although H-free carbene-like zigzag edges [15] and H-free dangling σ-bond zigzag edges [16] have also been proposed to explain magnetism in carbon materials, it seems unlikely that those edge sites can survive under ambient conditions (room temperature in air) due to their high chemical reactivity. The zigzag state observed in air by STS at room temperature has been attributed to zigzag edges with saturated σ-bonds [17].

To date, almost all theoretical and computational studies of graphene zigzag edges have focused on the physical aspects, such as electronic structures and magnetic properties. For example, the theoretical studies calculating weak ferromagnetism along one edge, and antiferromagnetism between two edges (one edge spin-up, the other spin-down) on a zigzag-edged graphene nanoribbon (ZGNR) [18].

From a practical perspective, the localized state at the zigzag edge offers the most useful electronic properties to help us achieve our goal. Due to the nonbonding character of the localized state and the closeness of the flat band to the Fermi level, the zigzag edge sites should theoretically be similar to radicals. Jiang et al. [19] have shown calculations on

GNR edges on how the edges distribute π-electrons and how these π-electrons contribute to reactions by external chemical stimuli. To show that ZGNRs are indeed unique (with respect to their useful surface area being the edge sites), they compared ZGNRs with a planar graphene sheet, nanotubes, and an armchair edge, for their reactivity with atomic hydrogen. They have shown that spin-polarized π-electrons are localized on the zigzag carbon atoms, which, from a chemistry viewpoint, prompts us to think of them as a "partial radical." That is, these ZGNRs have unpaired π-electrons distributed mainly on the two edges, but on average each edge carbon atom has only 0.14 electrons. Due to the partial radical character, these edge carbon atoms offer special chemical reactivity, compared to planar carbon atoms, armchair carbon atoms, or nanotube carbon atoms. In the reaction of the zigzag edge with a hydrogen atom, the bond dissociation energy (BDE) was calculated for the newly formed C—H bond in which C was an edge carbon [20]. The BDE changed only slightly for widths of $N = 4-6$. Anderson [20] also observed a ratio of zigzag faces to armchair was relatively consistent at 1.7:1 in platelet type GN. With this characterization, GN start looking even more practical for producing consistency in performance, capacities for functional groups, reactive groups, etc.

Comparing with other $C(sp^3)$-H BDEs (4.553 eV for C_2H_5-H and 4.315 eV for cyclo-C_6H_{11}-H) [21], the CH bond formed at the zigzag edge had a strength of ~60% of the C—H bond between a molecular carbon radical and H, again indicating a "partial radical" concept. The disparity of a partial charge of ~0.14 eV but the edge C—H bond generating a strength of ~60% of a common C—H bond is explained by another very important aspect of the chemical reactivity at the zigzag edges. Although the localized state at the zigzag edge offers only a partial amount of the π-electron density on a per edge carbon basis, remember these partial electrons are not confined to those edge carbons, but can act collectively when interacting with another radical [18]. So here the "localization" (or "localized" state) at the edge sites is meant to be with respect to the inner sites. Two pieces of evidence support the idea that zigzag edge π-electrons can respond collectively to an attacking radical. First, after the formation of a C—H bond at an edge coverage of 1/6, the local magnetic moments on the intact carbon atoms on the same edge greatly decrease. In other words the electronic states at the zigzag edge act together when a

C—H bond is formed at one of the edge carbon sites, even though the electron density is distributed evenly at the edge carbon atoms before the bonding.

The second piece of evidence is that the BDE is found to decrease with the edge coverage because there are fewer edge electrons available on a basis of per C—H bond formed when the coverage increases. Like the C—H bond, the C—OH and C—CH_3 bonds formed at the edge have a BDE of 50—70% of C_2H_5—OH and C_2H_5—CH_3 bonds [18]. For the halogens, the BDE of edge-X decreases from fluorine to iodine and follows the same trend as that of C_2H_5-X. This trend is attributed to the decreasing electronegativity between fluorine and iodine. Fluorine being the most electronegative has the greatest BDE. Moreover, one notes that the BDE ratio of edge-X to C_2H_5-X increases dramatically from chlorine to fluorine, indicating the extraordinary ability of fluorine to pull electrons from the zigzag edges.

By detailed analysis and calculations, it has been shown [18—20] that the zigzag-edged GNR has a significantly higher C—H BDE than the armchair-edged GNR, nanotubes, or a graphene sheet. ZGNR's unique electronic structure has a substantial peak near the Fermi level, which directly leads to its stronger bonding to hydrogen.

These collective research and experimental findings support the hypothesis that the chemical reactivity seen for the π-electrons of ZGNRs contribute to the chemical behaviors of carbon or graphitic materials which essentially have only edge sites. The implication here is that a uniform structure at the nanoscale with consistent platelet dimensions and possibly predictable bulk properties such as average zigzag to armchair face ratios could give quantifiable data to design and implement applications.

Finally, for those of you who like derivations in a scientifc discussion, Rotkin et al. [22] have proposed a four-parameter model toward a unified approach to define the energetics of the graphene edge regions. It has been shown [23, 24] that there are four approximately independent energy components, as well as a large constant contribution, which is the chemical potential of the carbon atom in a flat graphene. They counted the energy from the "zero level" of the infinite graphene sheet and omitted the large (~7 eV/atom) constant to arrive at the following, for the four contributions:

$$E = N_d \varepsilon_d + [N_6 / K(R_1, R_2)] * E_c + N_b E_b + N_{\text{cont}} W \tag{1.1}$$

where:

- N_d is the number of defects (primarily pentagons and heptagons) in the hexagonal lattice of graphite. It equals zero for the perfect graphite lattice.
- ε_d is the defect energy and consists, in general, of two parts: the dislocation core energy and the elastic energy accumulated in the distorted lattice around the defect. The latter is accounted for separately in the second term. This term is parameterized with the characteristic elastic energy:
- E_c (the elastic tensor, in general case) resembling the flexural rigidity of a thin shell [25–27]
- Any external stress should be included explicitly, but was not considered in the Rotkin model.
- K (R_1, R_2) is the tensor of curvature, and, with R_1 and R_2 being the principal radii of the curvature [28] defines the internal strain.

Therefore the whole second term diminishes to zero for the perfectly flat graphite structure. This term collects all elastic contributions and counts all atoms in hexagons, N_6, except the ones belonging to the defects or to the system edges. The surface energy (third term) is modeled at the microscopic level by counting the number of dangling bonds, N_b, multiplied by the dangling bond energy, E_b. This characteristic energy can vary with the change in the surface bond population.

The fourth term (van der Waals energy) is roughly proportional to the number of neighbor atoms in the adjacent layers, N_{cont}. Graphite, its modifications, and other layered crystals are described within the van der Waals theory in respect to the interlayer interaction energy W [15,29, 30–32]. These interaction forces are essentially short ranged, which justifies the use of the simplest approximation of the contact interaction.

The natural graphite flake has a hexagonal shape with the edges in [110] face and symmetry equivalent zigzag directions [16, 33]. The disoriented graphene edge can be considered as a series of kinks along the high symmetry direction (see Fig. 1.5-A).

Consequently, the surface energy may be calculated based on the disorientation angle. Similar to the free energy of 3D crystals [34], it has a cusp around the zigzag direction (minimum of the surface energy) and a parabolic shape around the armchair direction (maximum of the surface energy), which is shown in Fig. 1.5-B. A closed expression for the surface energy of the graphite edge with an arbitrary orientation was presented in Ref. [35]. It is based on the assumption that the local hybridization at the

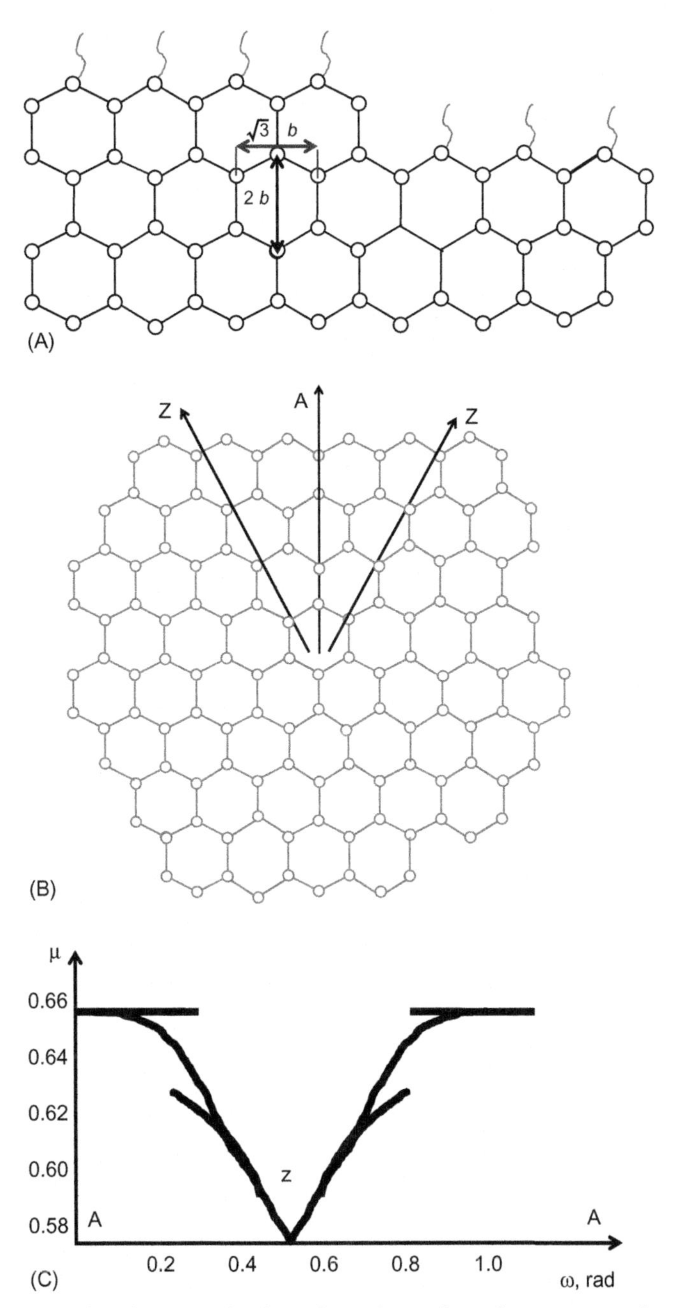

Figure 1.5 (A) The disoriented edge of graphite. The edge consists of a series of kinks. (B) The view of a small graphite flake, which shows that no continuum edge geometry is possible at the nanoscale, and a circular flake accepts a polygonal shape. (C) Dangling bond density, ν, of the graphene edge versus chirality angle, ω. The surface energy of the graphene fragment is a periodic function of ω because of $\nu(\omega)$. Horizontal lines represent the linear approximations at the extremum points: A, armchair edge; Z, zigzag edge. From: On Surface Energy of Graphene and Carbon Nanoclusters - S.V. Rotkin - 2001.

edge does not vary with the edge orientation. An almost free graphite edge is addressed, which may be passivated. The typical energy gain due to one kink creation is approximately equal to the half of the dangling bond energy, $E_b/2 \sim 1.2$ eV. This allows us to imagine the magnitude and the scale of energy needed to form the minimal disorientation of the edge. The formation energy of the disoriented edge is quite large compared to the synthesis temperature and, therefore, only zigzag edges at equilibrium should be considered. To minimize energy level then, a continuum planar flake will accept the circular shape. In reality, the density of bonds along the circular perimeter will oscillate (as shown in Fig. 1.5-C) and will increase its contribution to the total energy given by Eq. (1.1).

1.3 SUMMARY

A crucial factor in our discussion of applications with GN is the availability of large external surface area populated with highly reactive edge sites. I may even sound irritatingly redundant about this characteristic throughout the book, so bear with me. Virtually the entire external available surface area of GN is composed of dangling bonds associated with edge sites. Two types of edge sites exist, armchair and zigzag. The zigzag edges are more reactive to chemical attack. The highly reactive surface area with virtually no basal planes puts GN at an advantage for applications where a reaction must take place between the carbonaceous nanomaterial and a reactant. This could be for functionalization, in situ polymerization, or catalyst support. Other materials, including graphite, with large basal plane surface areas, require harsh oxidation procedures to create defects on the basal planes which resemble the edge sites inherently present in GN. So we can expect processing GN to make them ready for industrial use is therefore minimal.

CHAPTER TWO

Review of Carbonaceous Nanomaterials and Graphite

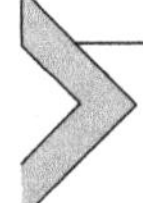

2.1 GRAPHITE

2.1.1 Background

We can now take a deeper look at the remarkable qualities imputed in graphite's behavior by it's unique electronic configuration. The semblance of "scientific discovery" of these core characteristics is in reality a confirmation of theoretical predictions and hypotheses.

Natural graphite is formed as a result of the dissociation of sedimentary carbon compounds during metamorphism. It is found naturally in three different forms: lump or vein graphite, crystalline flake, and amorphous. It's applications and uses are based on maximum benefits of its physical form.

Vein or lump: Only mined in Sri Lanka, it is the highest quality of natural graphite. It typically has a purity of 95–99% carbon as mined. It exhibits higher thermal and electrical conductivities, but more importantly, it mostly does not require polymer binders in order to be formed into solid shapes. This quality makes it suitable for high-purity graphite parts.

Crystalline graphite (CG): Crystalline graphite has a distinctive plate-like formation and is also called flake graphite. CG can be found in all sizes with purities of up to 95%. This type of graphite represents ~45–50% of the total graphite consumption today, and is the primary one we are concerned about for the purpose of this book.

Amorphous graphite: As the name suggests, it is made of very fine particles, with low graphite content. It can be used for low-end applications such as pencils. Amorphous graphite also has a large market and represents approximately the same consumption and market size as CG.

Highly ordered (or oriented) pyrolytic graphite (HOPG) refers to synthetic graphite. Such graphite is produced with templates such as silica zeolites and other ordered materials which can vaporize at high temperatures and leave a graphitic pyrolytic product behind. The qualification of this type

Graphitic Nanofibers.
DOI: http://dx.doi.org/10.1016/B978-0-323-51104-9.00002-9

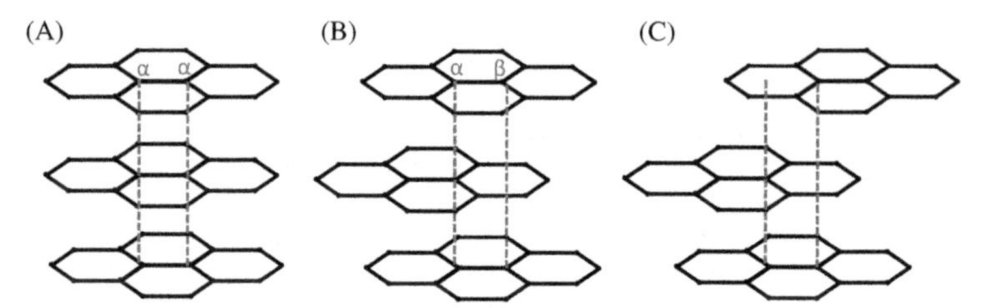

Figure 2.1 Stacking configuration of graphene planes. (A) AAA—rare—noticed, (B) ABA—85% of graphite, and (C) rhombohedral 15%.

of graphite is to have an angular spread between the graphene sheets of less than 1 degree. Expensive to make, it finds applications where consistent pore structure, stacking alignment (discussed below), and predictable properties are required.

From the basic discussion in chapter 1, essentially, the crystalline flake form of graphite, as mentioned earlier, is simply a total of millions of individual layers of linked carbon atoms stacked together. The stacks in natural graphite are typically aligned in the ABA (Bernal, 85%) or ABC (rhombohedral, 15%) (Fig. 2.1).

2.1.2 New bonding

Lets look at the actual energy levels of reactions on the non-edge sites of graphite. I have stated conceptually in chapter 1 that on-plane additions, defects, and reactivity conditions are difficult to control and predict. During chemical reactions, many structures can emerge. This is one of the disadvantages of using graphite to produce graphene oxide (GO). The energy levels of the bonds formed between the graphite plane and any additional carbon atom vary with their location in this hexagonal structure. On a singular plane, primarily the following configurations of bonds may occur. Li et al. [36] have calculated relevant values of bond distances and energies. Their work relevant to our discussions, with respect to (only single layer) new atom addition, is presented in Figs. 2.2-2.4.

2.1.3 Oxidation and functionalization

As we reviewed in chapter 1, an important characteristic of graphite is that a flake terminates with incomplete lattices, resulting in available, highly reactive edges. The selective functionalization of graphene edges is driven by the chemical reactivity of its carbon atoms. The chemical reactivity of an edge, as an interruption of the honeycomb lattice of

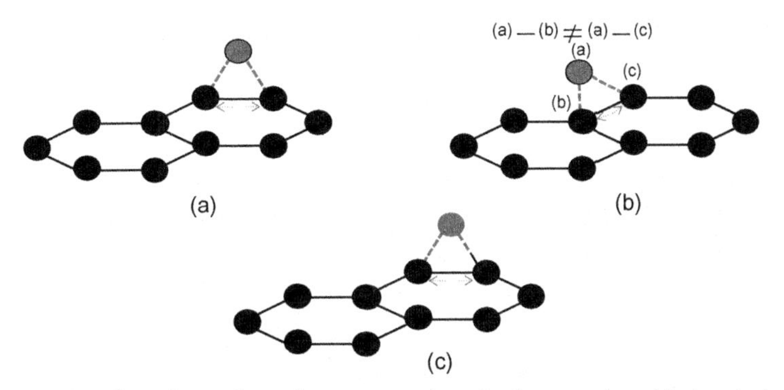

Figure 2.2 Bond configuration when new carbon in the new bond is located above the center of two planar bonded carbons.

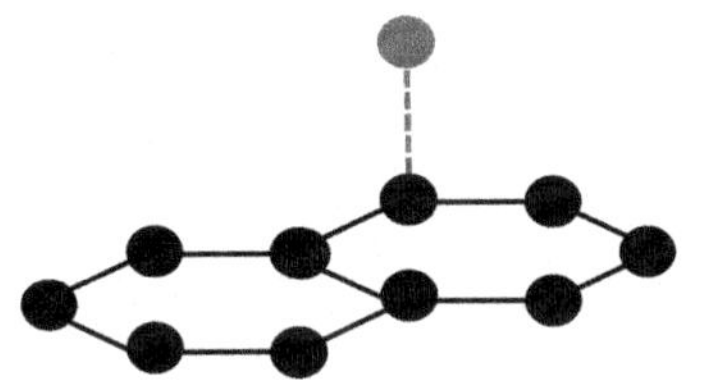

Figure 2.3 Bond configuration when new carbon in the new bond is located directly above another carbon.

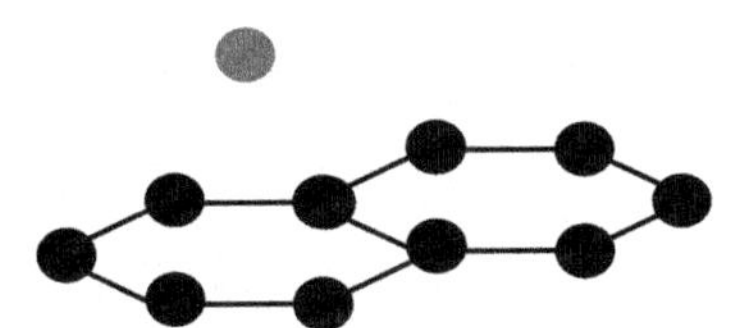

Figure 2.4 No bonds when new carbon in the new bond is located above the center of the hexagon.

graphene, differs from the relative inertness of the basal plane. However, when the sizes of the different basal planes are different, as in natural graphite, the dangling orbitals at the edges form π-bonds with some of the orbitals in the neighboring basal planes to reach minimum energy levels.

Oxidation of graphite and the functional groups resulting from various oxidation procedures is an important aspect of graphite and graphitic nanofibers and their possible applications. A quick review of the process and mechanisms with respect to basal planes and defect sites/edges will be helpful in understanding some of the applications discussed later.

The oxidation of carbon changes its physiochemical properties, the most significant one being that it can be made hydrophilic. Oxidation can also be used to tailor surface characteristics.

In the chemisorption of oxygen on graphitic/carbon planes, (spin adjusted) density functional theory (DFT) calculations show the absence of any kind of charge transfer between O_2 and graphitic basal planes [37]. This is not surprising, since much of the basal plane is at much lower energy levels than the edges are. The implication here is that the oxygen interacts with basal plane carbon with van der Waals forces. The reason for this nonreactivity is the energy mismatch of a few tenths of an electron volt between the unoccupied states of O_2 and the valence band of graphite. This necessitates a transformation of the unoccupied oxygen state for any reaction between oxygen and the graphitic surface to occur. Here the defect sites on the basal plane and the zigzag edges with the vacancies tend to bond with the unoccupied electron states in oxygen.

Two independent reaction mechanisms have been proposed for the oxidation of graphitic surfaces [36]: (1) reaction from the direct collisions of O_2 molecules with the reactive carbon sites (Eley–Rideal (ER) mechanism) and (2) the surface migration mechanism, i.e., the oxygen molecules are first adsorbed on nonreactive sites (Langmuir–Hinshelwood (LH) mechanism). The ER mechanism is directly initiated by the collision of the oxygen molecules with the graphitic defect sites [38]. The reaction mechanism according to LH consists of two distinct steps: (i) the physisorption of molecular oxygen on the graphitic basal surfaces. This interaction of molecular oxygen with a basal plane of graphite is physical in nature, without any charge-transfer; (ii) surface diffusion of the physisorbed molecular oxygen to vacancies (defects on the basal plane) or edge sites. Where it then undergoes a reduction to form species such as $(O_2)^-$, or $(O_2)^{2-}$ and finally makes stable covalent bonds with the carbon atoms.

Oxidation of graphitic nanofibers (GN), on the other hand, with entire surface area composed of edge sites, does not necessarily require physisorption and involves only the ER mechanism, i.e., the oxidation of GN happens by the direct collision of an oxygen molecule with the reactive edges of the graphene sheet [38]. Charge transfer from the carbon atoms has a reducing effect on the oxygen molecules approaching these sites. The resulting repulsion within O_2 breaks the O–O bond exothermically without an activation barrier and results in the formation of carbon–oxygen functional groups. GN can be oxidized with H_2O_2 or mild acids very effectively, resulting in –COOH and –OH functionalities at the edge sites. H_2O_2 was used to oxidize GN and decomposed to oxygen and water [37]. This catalytic activity of the edge sites in graphite or GN is indeed used in the water treatment industry to oxidize H_2S, which we will discuss later. HOOH decomposition in the absence of the edge sites would require a catalyst.

In gas-phase oxidation, below 600°C, oxidation is initiated by the chemisorption eventually on to the defect sites. Above 700°C, graphite can be oxidized by the etching of carbon atoms from the basal plane, resulting in dangling bonds. [39]. Thermal oxidation, i.e., heating the graphite surface at temperatures typically between 600°C and 1000°C in an oxygen partial pressure between 10^{-4} and 10^2 mBar, is the most common technique for the oxidation of graphite [40]. Surface carbon atoms are depleted without control in the form of carbon monoxide and carbon dioxide. The exposure of the carbon surfaces to oxygen at temperatures above 700°C also results in undesired surface groups, and further reactions between them, making the production of a consistent quality product even more elusive [41].

Typical groups observed on oxidized carbon surfaces are given in Fig. 2.5.

The profusion of functional groups formed on natural carbon surfaces is due to the varied reactivity at the different types of sites, partially because of impurities in the raw material. These impurities can act as catalysts or rate-limiting factors. Another reason is the unaligned nature of the edge sites, as per the stacking discussed above. Except for AAA stacking, which is rare, there is always a portion of the basal plane exposed to the oxidation process. Many organic species with unknown mechanisms have been reported to get attached to the basal planes. The delocalized state of the p-orbitals make the basal planes react via π-bonding in many

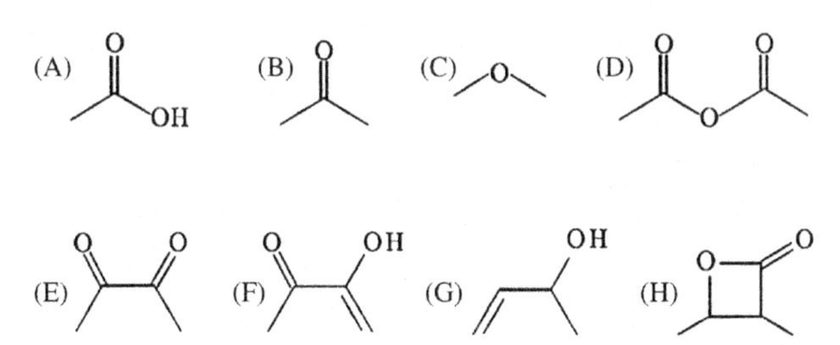

Figure 2.5 (A) Carboxylic acid, (B) keto, (C) ether, (D) anhydride, (E) quinone, (F) phenolic, (G) hydroquinone, or (H) lactone. *Reproduced from R. Zacharia, Oxidation of graphitic surfaces (Dissertation online), Freie Universität Berlin (Chapter 5), May 2004.*

different energy states to form organic complexes. Many of the organic groups undergo secondary reactions with each other after attaching to the basal surfaces to further form new compounds [42]. Due to this exposure, the surface area measured by the Brunauer Emmett Teller (BET) technique for either ABA or ABC configuration is more than the surface area measured for materials like GN that have only edge sites. With a combined exposure of different levels of energy (basal vs edge), it is not difficult to imagine the complex nature and range of electronic perturbations in these regions. One can imagine the arduous task of delienating a synthesis protocol, that could be implemented to produce a product with predictable performance characteristics, be it graphite, nanographite, nanographite flakes, graphite nanoplatelets, multilayer graphene, or any other fancy nomenclature assigned to natural graphite with nano dimensions. Industry needs to have reproduceable, consistent, and predictable raw materials for successful commercialization of their downstream products. One important parameter desired would be the ability to load a narrow range of moeties or equivalent weights of functional groups per unit of mass of the graphitic material. If we take away the exposed basal planes from the picture, the edge states, having only two possible configurations, are more confined in their electronic behavior and offer more predictable bonding possibilities.

It would stand to reason, then, that synthesized graphitic nanomaterials such as GN with minimal exposed basal surface area, coupled with a relatively narrow range of ratios between the two types of edge states, may offer vast possibilities for commercial practicality. The functional groups, the protocols to attach them, etc., could become standardized procedures in industry. We could then formulate functions, capacities, and operating conditions to actually achieve a big portion of the hyped superior performance of products.

Carbon–oxygen surface functional groups determine the many physicochemical properties of the carbon surfaces, such as surface acidity, wetting, etc. These surface concentrations of the carbon–oxygen groups are temperature-dependent. Their ease of formation and stability depends on the materials, methods, and temperatures of preparation. Developing industrial scale manufacturing techniques, that produce a narrow range of capacities of particular functional group equivalents per gram, would have tremendous implications in various fields such as catalysis, wastewater treatment, etc. Significant progress has been made in identifying the types of functional groups, their location on the platelet, their stability, etc. [42,43]. We will look at this aspect again later.

Thermal stability of the surface functional groups is also important for identification, purification and quality control. A technique known as thermal desorption spectroscopy (TDS) is often used to achieve this objective. In TDS the temperatures at which carbon dioxide and carbon monoxide desorb from the surface are compared with data derived from decarbonylation and decarboxylation reactions (i.e., the elimination of CO and CO_2) using classical pyrolytic elimination reactions. The classically derived values given in Table 2.1 are from various sources, and are subject to inaccuracies given they were derived from different carbon types, with different chemical pretreatments, etc. Nevertheless, they give a reasonable method for identification of functional groups by desorption.

The variation in functional groups formed during oxidation of graphitic materials is a critical feature that can be exploited in the functionalization of these materials. (more on this later). Modern technologies help us accurately gain information about the composition, electronic state, chemical state, binding energy, layer thickness of the surface region of solids, etc. A technique called Photo Electron Microscopy (PES) is also used. The two variants within this technique are not relevant here, because they can both provide data we are concerned with. The sample is irradiated with a beam from a high energy source (such as X-Rays or ultraviolet rays). An analyzer then measures the number of electrons that escape from a very thin surface layer of single digit nanometer heights, as well as their kinetic energy to produce a spectrum. This spectrum is a graph of emission intensity vs binding energy. The elements on the surface are thus accurately identified based on the unique binding energy nature has bestowed upon each element. The peak areas on these spectra can also be used to obtain the concentration of the elements. The groups

Table 2.1 List of surface oxygen groups on oxidized graphite

Surface group	Product	Temperature (°C)
Carboxylic acids	CO_2	180–300
Acid anhydrides	CO, CO_2	400–447
		437–657
	CO_2	137–157
		190–650
Peroxides	CO_2	550–600
Phenol	CO	600–700
Quinones	CO	800–900

Reproduced from R. Zacharia, Oxidation of graphitic surfaces (Dissertation online), Freie Universität Berlin (Chapter 5), May 2004.

on edges have been identified by experiments. Martin Rosillo et al. [42] investigated the functionalization of pristine graphene edges obtained from nanotubes via PES. The graphene edges showed no elements other than carbon and oxygen. A high-resolution spectrum of the region confirmed the presence of sp^2 carbon with a peak centered at 285.3 eV as well as a lower intensity peak at 289.4 eV which is attributed to carbon in oxidation state +III such as in carboxyl (−COOH) groups [29,33]. Comparison with the same region of GO prepared by the Hummer's method showed that the most pronounced peak was located at 287.2 eV, which would confirm alcohol and epoxide groups on the basal plane [29]. −COOH groups are present in very small quantities at a spectral intensity around 290 eV. On the other hand, for GN, XPS comparisons indicated the presence of sp^2 carbon and carboxylic groups, and absence of any alcohol and epoxide groups as seen with GO basal plane oxidation. This sets the stage for predicting −OH and −COOH functionalities on the edge site surface area of materials with minimal basal plane surface area, such as GN.

Another experimental comparison was done with TDS to study the differences in features of oxygen from edge zones and those from colloidal graphite [37]. This effort also reveals that in graphite, oxygen also binds to a low binding energy basal plane. The reaction therefore seemingly proceeds via the two-step reaction mechanism described before (L-H). The oxidation of the edge sites follows the ER mechanism, i.e., a one-step oxidation with direct collision of O_2 molecules. The charge transfer and the formation of carbon−oxygen functional groups by the single-step mechanism (ER) can be envisioned similar to reactions of polyaromatic hydrocarbons (PAHs). The justification for this comparison here, is the structural and electronic similarity between PAHs and graphene sheets. In this example the oxidation of PAHs by hydrogen peroxide to produce quinones is used for illustration (Fig. 2.6). The single-step reaction between oxygen molecule and carbon atoms is assumed to proceed

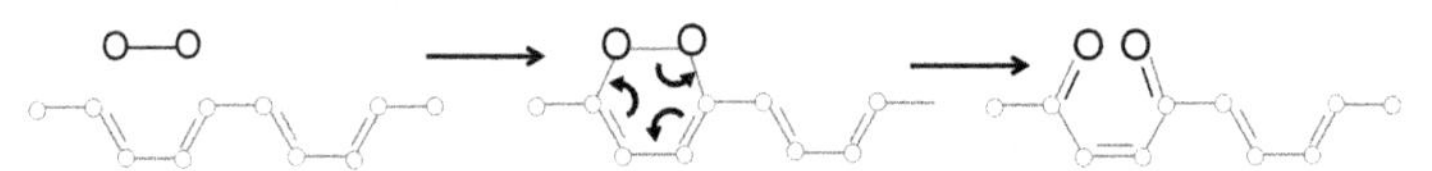

Figure 2.6 The oxidation of PAHs by hydrogen peroxide. *PAHs*, polyaromatic hydrocarbons. *Reproduced from G.J. Gleicher, Hydrogen abstraction from substituted benzyl chlorides by the trichloromethyl radical, Tetrahedron 30 (8), (1974) 935−938 [44] referenced by R. Zacharia, Oxidation of graphitic surfaces (Dissertation online), Freie Universität Berlin (Chapter 5), May 2004.*

through two elementary steps as shown: The first step (1) involves the chemisorption of oxygen molecule on the armchair, which results in the formation of an aromatic peroxide surface functional group (2). This peroxide surface group undergoes a rearrangement and leads to the formation of a 1,4-quinoid surface functional group (3) [43].

H_2O_2 (l) and O_3 (g) as oxidizers on fibers with smooth walls have been used and some defects have been investigated [43]. Given our focus on applications, we can draw a logical parallel from this work to emphasize the importance of edge sites versus basal (defect) sites. Rosenthal's [43] results undoubtedly show that the final nature of the functional group was dependent on the nature of the surface. They oxidized carbon fibers with some graphitic content. At defect sites (sharp edges), the functional groups were determined to consist mainly of –OH and –COOH moieties. Some surfaces which lost their edges, defects (termed graphitization by the authors) were found to anchor epoxides, ketones, and other complex structures. In addition, they found O_3-oxidized edges to retain the edge states and result in higher carboxylic and hydroxyl functionalities. In the characterization section for GN, I will expand on this knowledge with further experimental data of the quantum of edge sites, their proportions, etc.

2.1.4 Summary

Graphite is a material with very unique (and well-known) characteristics. These qualities do not change with particle size, until we get down to 1- to 3-layer graphene flakes with lateral dimensions between 5 and 10 nM (graphene quantum dots). At that point the electronic properties are affected by the size of the basal plane as well as the nature of neighboring orbitals. From the experience gained by researchers, it is possible to extrapolate or hypothesize many applications with graphitic materials. As mined, natural graphite has uneven basal planes making up for most of the surface area. The varying contaminants on these planes create different kinds of defects when they are oxidized, forming a myriad of complex organic species, which have not yet been completely understood or characterized. Most importantly, the desired properties of any carbon/graphitic nanomaterial are derived from the highly reactive edge sites. When the edge site has another flat edge atom neighbor it is termed the armchair face, while the edge with three coordinated bulk lattice neighbors is termed the zigzag face.

In chemistry applications the utility of a material relies heavily on the types of functional groups that may be added with predictable capacity on the material. Oxidation is the most popular route to functionalize graphitic planes. Oxidation generally occurs at the dangling bonds existing on sharp edges or on the basal plane via physical sorption and subsequent surface migration of the oxygen to the defects in basal planes. To increase reactivity for the planar smooth surfaces of natural graphite, basal planes, "defects" are created on the basal planes by oxidation to create the dangling bonds of an empty sp^2 orbital and have a reducible oxygen compound so that a more active zone is formed. The different energy levels and subsequent reaction mechanisms generate a variety of organic compounds with no predictable compositions. The edge sites of graphite, on the other hand, oxidize in one-step reactions resulting from direct collision with molecular oxygen, liquid or gas phase.

Between the two types of edge sites, the importance of the edge p-orbitals at the zigzag sites has been established as critical for functionalization, and if the only available active surface area in our material constituted edge sites, we would welcome the absence of the basal planes for most of the practical applications. Conventional surface area measurements typically include the edges as well as the basal planes. So the only surface area that matters for people in the commercial and industrial world is the surface area offered by the edge sites. This surface area can only be quantified for useful purposes in GN. By nature of the raw material, graphite and GO are not amenable to be reproduced with predictable surface areas of either type; exposed basal or edge.

2.2 GRAPHENE

Graphene is the thinnest material we know of and the strongest ever measured. It is sufficiently isolated from its environment to be considered free standing. Yet it is not found in single-sheet form in nature. It probably is also the most misunderstood product in the commercial world today. While the definition is quite elementary, (a single, 2D layer made up of octagonal arrangement of carbon atoms), it has been abused by marketing departments of many companies. If it is not a single layer, it cannot technically be called "Graphene." But commercial convention accepts 5- to 10-layer graphite as "graphene." If it does not fall in this range, it is called Few Layer Graphene (FLG), but the word "graphene" is a must in the definition, or there is no value to the glorified graphite product or research publication. To the chagrin of scientists who genuinely can open new frontiers in the understanding of particle physics, quantum mechanics and improve other atomic level knowledge with single layer graphene experiments, the world is approaching "graphene fatigue". Unfortunately the field has gained hype in recent years in the investment circles, where decision makers are guided by a camarilla instead of people with a sound understanding of the science, practicality of manufacture, and economic viability.

Due to the unpaired p-orbital, charge carriers in pristine (defect free) graphene exhibit giant intrinsic mobility, have the smallest effective mass (zero), and travel μ lengths without scattering. Graphene is impermeable to gases, including helium [45]. Electron waves through the graphene lattice result in quasiparticles and are characterized by a Dirac like equation, which marries quantum mechanics principles to the special theory of relativity and allows us to derive the low lying (core) states of heavy atoms, where the speed of electrons close to the nucleus approaches the speed of light.

To date, the only type of graphene that has benefited science, has been the crystalline, pristine, and defect-free form. It is still mainly produced by the original way it was discovered, i.e., physical peeling by adhesive tape [46]. The quantity produced seems to be able to satisfy current demand. Large flake, single-layer graphene in pristine condition have not yet been synthesized in a practical manner. To be sure, almost all the applications in the large, commoditized markets such as plastics that are being hyped or promoted by the word "graphene" have no commercial or practical merit outside of the lab environment today. There is no defined mass production method, quality control protocol or any semblance of affordability. Furthermore, many of those applications, such as improving thermal and electrical conductivity, would not really need such a high-quality material to achieve the objective. Without a doubt, what can technically be termed as graphene, and its properties are helping fundamental research in physics in unprecedented ways. For most of us, this should be the silver lining in this confusion. For example, Klein tunneling has been experimentally predicted using the Dirac equation. Other theoretical phenomena such as the hypothetical rapid motion of elementary particles, (zitterbewegung) and the nonperturbative effect particle production from a quantum vacuum (Schwinger) production can now be possible. With existing tools such as scanning probe microscopy, graphene can enable detection of local magnetic moments, mapping of wave functions in quantizing fields and other phenomena and properties that our esteemed (true) researchers can elucidate for us in simple terms, guiding us in the future to develop products with better performance. The operative word is "future". The discoveries are in the embryonic stage and no one to my knowledge is in a position to forecast time frames of how these discoveries can be utilized for consumer use, beyond headline grabbing speculations. These headlines are fodder for unrealistic presentations to investors and consumers. Therefore the most outstanding benefit of the discovery that single sheets of graphene can be isolated is that its remarkable qualities will help physicists and chemists explore new frontiers, which will lead to new discoveries in materials science. Commercially, I suspect the research market will dominate as the major consumer of defect-free graphene flakes, and a small demand will be generated by high-end applications that require very small size and very high purity [46–74]. Graphene does have immediate applications.

1. *Transmission electron microscopy (TEM)*: TEM is a technique to study material properties by shooting a beam of electrons at an ultrathin specimen of the material. Images can be formed based on the interaction of the electrons transmitted through the material sample. The sample is supported with the supports that generate signal and noise contributed by sample charging and instability. Graphene, with its single atom thickness, can provide extreme physical stability, periodic structure, and ballistic electrical conductivity.
2. Nanoelectro mechanical systems (NEMS) [75–78] can use graphene properties for sensing applications. Graphene-based resonators with ultrahigh frequency will enable NEMS to achieve the inertial sensing of individual atoms and the detection of zero-point oscillations.

These are niche applications and even at a mature market state, will create a very small demand for graphene. For some applications, graphene is a great candidate and maybe few-layer graphene (FLG) is very close to being qualified for use, such as individual ultrahigh-frequency analog transistors. GaAs-based high electron mobility transistors (HEMT) can be replaced in communication technologies by graphene, though there are certain limitations, but graphene offers a possibility to extend HEMT's operational range into terahertz frequencies.

One can easily see why these realistic applications for graphene are not shown in the promotional endeavors of graphene start-ups. These are what one would consider niche markets, with very low volume, but in a position to pay high prices. A researcher is happy with 5 g of graphene flakes, while business plans are being presented for thousands of tons per year for unrealistic applications. Hundreds of millions of dollars have been invested into companies promoting "future" applications of graphene, but presenting no viable roadmap on how they intend to produce graphene with the consistency and at a cost that is necessitated by commercial viability requirements. Graphene is not a material that is easily extracted/synthesized, so projecting widespread use in commoditized (high-volume, price-competitive) materials is simply naïve, and/or misleading depending on the intent. If one studies the commoditized materials market (such as polymers, lubricants, drilling fluids, catalysis, etc.), very few large volume applications are in a state such that they face extinction unless graphene was to come and save the day. None come to mind. There is always a more expensive plastic, rubber, lubricant, etc., available in the present at a much lower cost increment, that performs sufficiently well for higher performance market needs.

Some large-scale applications show promise of high-volume markets, but without a roadmap for mass production of graphene, those applications are so far off in the future, even the top of the mast is not visible yet on the horizon, let alone a boat coming in. For example, theoretically, graphene is a viable candidate for the replacement of indium tin oxide, most commonly used in photovoltaic cells and displays. To date, the "candidate" optoelectronic coatings have exhibited unacceptable levels of resistance (100 times higher) for the required transparency in such cells [79–83]. Even more exotic methods are probably being promoted out there that I am incognizant of, especially in the sexy field of photovoltaic power generation. I seriously doubt any one can convince me that a single layer graphene wafer can be cheaper than single crystal silicon on a $ per watt basis. Cost is often obfuscated by dazzling efficiency numbers, but the market eventually asks the same question, how much will it cost me per watt of power generated?

Another commonly quoted application is the replacement of Si in integrated circuits as the base material. The hype seems oblivious of the hurdles such as graphene's gap-less spectrum. In addition, from a commercial perspective, tools and methods to define the structures at the atomic level outside the laboratory environment do not exist. Quality control, consistency in performance, etc., at the atomic level are not to be taken lightly. Admittedly, these applications may become reality in the future, but in 2016, much research and experimentation is still due before a commercial path can even be envisioned.

Methods used on a laboratory scale have been promoted as roadmaps to low-cost large-scale graphene synthesis. Mostly they turn out to be attention grabbing headlines, with no explanation of how much it will cost, or the practicality of the manufacturing process [84–89]. There is a prolific amount of literature that uses the GO route to make FLG, but the produced graphene cannot be made defect-free. The GO intermediate is made by the harshly oxidizing Hummer's method. I will outline the cost of manufacturing this material and others in Chapter 5, Cost of Manufacturing.

2.2.1 Summary

Graphene is a material, that offers us a window into a world we can only imagine. Its properties will provide insight into hitherto inaccessible levels of atomic structure behaviors. It is the backbone and the main ingredient in nanotubes and other nanomaterials. Large-scale manufacturing of graphene, being touted by many companies will hardly be justified for very simple reasons. In bulk applications, simply adding crystalline GN can achieve overall bulk performance improvements at much better price to benefit ratios [90–94]. However the implicit ceiling for cost of manufacturing and sustainability of supply of any carbonaceous nanomaterial will still be set by the price of high-quality graphite, since graphite can achieve a high percentage of the functions the nanomaterials are being promoted for. There are many applications that already use graphite with great success, such as thermally conductive lubes, adhesives, etc. A cost-to-benefit model would be easy to derive from those markets. Applications that can really capitalize on the unique electronic and mechanical properties of the pristine graphene plane, such as electronic tunneling, short-distance resistance-free current flows, etc., are not going to be very high-volume consumers. Nevertheless the economical isolation of this basic plane in graphite is a worthwhile technical goal to enable new materials and enhance existing ones, such as semiconductors, etc. The scientific exploration and research by students with fecund imaginations must continue, but laboratory-scale experimental results claiming potential profits in the short run, must be taken with extreme caution.

2.3 CARBON NANOTUBES

Before there was graphene, there were carbon nanotubes (CNTs). They could (theoretically) facilitate elevators to the moon, bulletproof apparel as thin as silk, and enough flat format electronics integrated in your shirt to act as data centers for all your needs, including transmitting real-time physiological data to your doctor. After millions, maybe billions of dollars, reality struck home. Not a single parameter required for deployment has been achieved at a reasonable cost and simplicity. However, CNTs do form a very important and unique class of nanomaterials, and much research on the benefits of CNTs (without misnomers) has been accomplished. This research can be used to draw parallels between GN and CNTs to justify the use of the former in most of those applications. Hence a discussion of their nature and properties will help in my endeavor to justify my thought of "good enough" achievements for industrial products.

CNTs are a result of the strong sp^2 hybridized carbon–carbon bond and exhibit high electrical and thermal conductivities as well as high mechanical strength. CNTs can be imagined as rolled graphene sheets that form long concentric cylinders (while graphite is made of stacked graphene sheets). There are two main types of CNTs: multiwall (MWCNTs, I include Double Wall (DWCNT) in this category) and single-wall (SWCNTs). For purposes of this introduction, both are sufficiently described jointly.

MWCNTs are multiple concentric SWCNTs. Often capped at both ends, with diameters of several nanometers to 200 nm. Van der Waals forces hold these concentric tubes together, like they do the basal planes in graphite.

2.3.1 CNT configurations

There are three configurations of CNTs: (1) zigzag, (2) armchair, and (3) chiral. The zigzag structure is one where the hexagons are lined up around the circumference of the nanotube (Fig. 2.7A). The armchair structure is one where the hexagons are lined up parallel to the axis of the nanotube (Fig. 2.7B), and the chiral structure is one when there is no particular alignment and the nanotube seems to be twisted, at an angle termed the chiral angle (Fig. 2.7C). The structure of chiral CNTs may be described in terms of the tube chirality, which is defined by a chiral vector

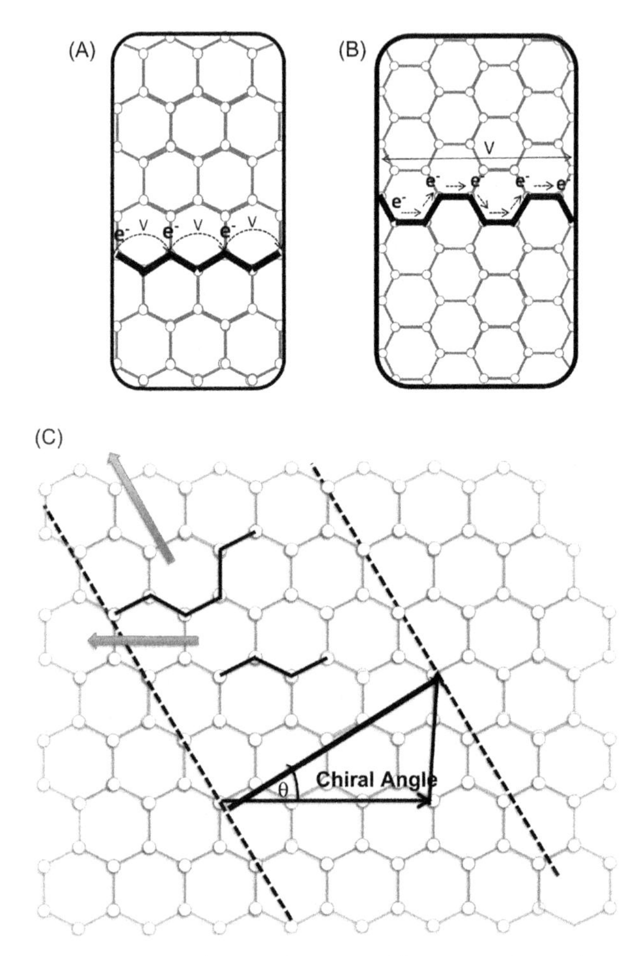

Figure 2.7 Chiral structure of nanotubes: by rolling a graphene sheet in different directions, typical nanotubes can be obtained: (A) zigzag, (B) armchair, and (C) chiral. *Adapted from M. Kole, T.K. Dey, Investigation of thermal conductivity, viscosity, and electrical conductivity of graphene based nanofluids. J. Appl. Phys. 113 (2013) 084307.*

and the chiral angle θ (Fig. 2.7C). The chiral vector indicates the way, in which graphene is rolled-up to form a nanotube:

The chirality of the CNTs has an impact on their properties, especially electronic ones [6].

CNTs (especially the walls) possess many properties similar to graphene. For commercial applications, they are expected to beat graphene in viability. But graphene has proven to be far more useful to the research community than CNTs have. And the research market is the largest segment for any synthesized carbon nanomaterials at present. After graphene became the darling of the industry, investments in CNT businesses were abandoned mid track, CNT manufacturers with the task of producing large-scale CNTs (for commercial applications) at a cost which is nowhere in sight. With much fanfare, Bayer had set up a manufacturing plant for CNTs in 2010, only to shut it down recently citing lack of market demand. One would think that a multinational such as Bayer would have the holding power and captive demand from their chemical and polymer corporate networks to justify waiting for the market to develop, for commoditized products if it was even vaguely in sight. I present a discussion of the properties CNTs to give insight into the viability of substituting other graphitic materials as candidates to achieve "good enough" performance for applications at a cost that is reasonable.

2.3.2 Electrical properties

The presence of mixed chirality in nanotubes with the currently available manufacturing techniques limits widespread adaptation of these materials contrary to what was initially assumed during the hype phase. One of the major characteristics this factor imparts to CNTs is that they can either be conducting or semiconducting.

2.3.2.1 Semiconductors

The two structures (zigzag and chiral) (Fig 2.7 A and C) do not conduct electricity unless energy equivalent to releasing an electron from the carbon is applied, such as a magnetic field or ultraviolet light. Such nanotubes have attracted much press as being possible replacements to silicon and GaAs semiconductors. In reality, these applications have a very long way to go to commercialization. With today's production know-how, the chirality of the produced nanotubes is mixed and varies. Controlling production to produce a specific type and diameter nanotube is indeed a challenge, and to my knowledge has not been achieved, at least not beyond laboratory efforts. In addition, for semiconductor applications, the geometry of the nanotubes determines band structures and thus the energy band gap. The energy band gap of semiconducting CNTs highly depends on the nanotube diameter and is given by:

$$\text{Energy}_{\text{gap}} = \frac{2\gamma_0 a_{C-C}}{d}$$

where γ_0 denotes the C–C tight binding overlap energy (2.45 eV), $a_{\text{C-C}}$ the nearest neighbor, C–C distance ($\sim$1.42 Å), and d is the diameter of a nanotube. There lies another hurdle. Very few processes can produce consistent diameter CNTs today.

2.3.2.2 Metallic conductors

The configuration where the hexagons are aligned to the axis of the nanotube is called the "armchair" configuration (Fig. 2.7B), and these nanotubes are metallic. Armchair structures can be more conductive than metal wires, when there is any potential across the ends.

Again, from a practical standpoint, it would be far more conceivable that this property will be the one to propel CNTs into the commercial world, if given a chance. Given their tubular structure, high aspect ratios (length divided by diameter), and agglomeration tendency, CNTs align themselves in a connected continuous phase imparting conducting properties to otherwise nonconducting materials. Small discontinuous particles such as graphene would have trouble meeting CNT performance in this area. The probability of agglomeration in graphene would not be high enough, and therefore, a larger percentage of graphene flakes by weight would be required as additives. The only form of GN that would give similar performance would be the ribbon type GN. The metallic properties of the MWNTs are due to their multiple-shell structure consisting of tubes with various electrical properties, where additional electronic coupling between shells takes place. Moreover the graphene surface of the MWNTs is predicted to have ballistic electron transport at room temperature (this term refers to conduction where Ohm's law does not apply; the resistance is not dependent on the CNT's length). The electrical current that could be passed through a multiwall nanotube corresponds to a current density in excess of 10^7 A/cm^2. The conductance for CNTs is given by:

$$G = G_0 M = \left(\frac{2e^2}{h}\right) M$$

where $G_0 = (2e^2/h) = (12.9\ \text{k}\Omega)^{-1}$ is the quantum unit of conductance, e is electron charge, h is Planck's constant, M is an apparent number of conducting channels which includes electron–electron coupling and intertube coupling effects in addition to intrinsic channels. The graphene planes in nanotubes are considered to provide one-dimensional electron flow. Experiments by researchers at Berkeley and Delft have shown that electrons can travel for long distances in nanotubes without being backscattered. This is in striking contrast to the behavior observed in traditional metals like copper, in which scattering lengths from lattice vibrations are typically only several nanometers at room temperature. The main reason for this remarkable difference is that an electron in a 1D system can only scatter by completely reversing its direction, whereas electrons in a 2D or

3D material can scatter by simply changing direction by a tiny angle. Phonons—long-wavelength lattice vibrations that scatter electrons in both 2D and 3D materials at room temperature—do not have enough momentum to reverse the direction of a speeding electron in a 1D nanotube. They therefore do not influence its conductance, at least not at low voltages.

Under ordinary conditions, a 2D or 3D metallic conductor behaves as a "Fermi liquid." Electrons in such materials fill the low-energy states up to the Fermi energy, creating what is well known as a "Fermi sea" of electrons. The low-energy excitations (or "quasiparticles") of this system act almost completely like free electrons, moving entirely independently of one another. In other words an excited state looks very much like a single extra electron above the Fermi sea. In 1D systems, on the other hand, the low-energy excitations are collective excitations of the entire electron system. The electrons move in concert, rather than as independent particles of a Fermi liquid. This system is referred to as a "Tomonaga–Luttinger liquid" (or, more simply, a Luttinger liquid) to emphasize its difference from the standard Fermi liquid behavior of 2D and 3D metals. Other electronic properties are discussed in many publications [95–103]. For other applications where high surface area and high surface energy are required, such as catalysis and functionalization, CNTs need to be reprocessed to have an etched (created defects) or stepped surface with more dangling bond energy. For example,

- Platelet fibers are deposited/grown on CNTs [104,105].
- CNTs are "etched" by various oxidation methods to create imperfections, which create larger surface area as well as higher energy states.

2.3.3 Purification

Besides the difficulty of producing the desired type of CNTs, another conundrum with CNTs is in the complexity of purifying the product to meet specifications that are required for applications. As-synthesized CNTs inevitably contain carbonaceous impurities and metal catalyst particles. Polyhedral carbons and graphitic particles that have a similar oxidation rate to CNTs, are the most difficult to remove. It is also difficult to remove metal impurities from the catalysts, which are usually encapsulated by carbon layers. Carbonaceous impurities typically include amorphous carbon, fullerenes, and carbon nanoparticles and can be removed with many processes, though due to physical issues such as agglomeration, none of them show a clear one-step approach.

- Solvents remove fullerenes, because they are soluble in many solvents.
- Liquid-phase oxidation: Amorphous carbon has a very high density of defects, so it can be removed by liquid oxidation. But this form of oxidation is very harsh and ends up damaging the walls of the CNTs as well. Indeed, chemistry similar to the Hummer's method is used to achieve this. The oxidants create defects on the surface, which then get attacked by oxygen groups in the solvent.
- Gas-phase oxidation: The gaseous oxidation method can effectively remove amorphous carbon (as CO_2 and CO), though it cannot remove the catalyst particles. For fluidized bed CVD reactors, gas-phase oxidation could be more convenient, but more difficult to control. Unlike liquid-phase oxidation, CNTs exposure to the oxidant is not evenly distributed, and while the amorphous carbon may get oxidized, a portion of the CNTs may be exposed to the oxidant longer than desired, and get damaged, or oxidized themselves. So better contact time (distribution) as well as some mechanism to reduce the oxidation rate of CNTs would be required. Park et al. [106] have reported gas-phase oxidation of MWCNTs without metal removal as well as suppressed oxidation by adding a reducing agent (H_2S) to the oxidizing gas. The reaction proposed is:

$$C(s) + H_2S(g) + O_2(g) = COS(g) + H_2O(g)$$

They achieved a yield of 54 wt% for MWCNTs. For SWCNTs, they removed the metals first in a preceding step by liquid-phase acid treatment. The authors do not elucidate the subsequent treatment of the carbon oxy sulfide (COS) formed during this process. Removing COS from the exiting gas stream is necessary to meet air emission standards, and the costs can be substantial.

Another way to increase the difference in oxidation rates between MWCNTs and amorphous carbon impurities would be by intercalation Chen et al. [107] report a combined purification process consisting of liquid-phase bromination followed by selective oxidation with oxygen at 530°C for 3 days. The brominated sample oxidized much better than the unoxidized sample. In addition, TEM studies showed the CNTs in the brominated samples after purification were open. They achieved a 10–20% yield. Fluidized Bed Chemical Vapor Deposition (FBCVD) reactors could be used for the oxidation cycle. However the oxidation cycle could not be carried out immediately following the synthesis cycle because the metal catalysts remaining in the CNTs would catalyze the oxidation reaction and destroy the CNTs [108,109]. If we need to remove the metals, an acid wash in the liquid phase would be required.

- *Thermal treatment*: Any of the above methods can be complemented with thermal annealing. At high temperatures, the metals would vaporize and the damaged CNTs walls would be repaired back to the crystalline state. Unfortunately, any amorphous carbon would also be converted to graphitic material.
- *Physical methods*: Physical means such as filtration and centrifugation separate CNTs from impurities based on the differences in their physical characteristics. In general, the physical method can be used to remove graphitic sheets, carbon nanospheres, aggregates or separate CNTs with different diameter/length ratios. Because these methods do not require oxidation, they do relatively less harm to the CNTs, especially the MWCNTs. As you can imagine, the density differences are not significant enough to simply do a conventioanl single stage centrifugation. The physical methods are always complicated and less effective.

The diversity of the as-prepared CNT samples, such as CNT type, morphology and structure, as well as impurity type and morphology, needs a skillful combination of different purification techniques to obtain CNTs with a particular set of characteristics and level of desired purity.

The discussion above defines the paradoxes associated with producing the desired type and purity of CNTs. Since most of the applications of CNTs will require pristine CNTs, the purification processes have to be combined and repeated several times currently, to make them suitable for the research topic mentioned in any publication. If your interest in the publications is beyond technical, prudence dictates experimental section of these publications should be studied carefully.

While CNTs have enjoyed considerable press and popularity among investors, the myriad of applications suggested have not moved beyond lab scale or small-scale production.

2.3.4 Summary

The relevant characteristics for our discussion later of GN:

1. The walls of CNTs are smooth, and with minimum defects. The resulting available charge density makes these tubes agglomerate.
2. Purification of these materials is one of the costliest components of the manufacturing process. The net yield is $\sim$10% of the feed carbon.

3. To create reactive sites (similar to the highly active edges of graphite or nanographite and GN), defect sites must be created by harsh physical, chemical, electromechanical, or chemical means.
4. CNTs synthesize as a group of desirable and undesirable types. Separating them and purifying them to the levels required for commercial applications is not yet been achieved on a commercial level.
5. Ribbon type GN would be a low cost alternative to achieve "good enough" dosage to performance ratios to that of CNT while other nanomaterials cannot replicate some of the CNT characteristics.
6. If work continues on CNTs, and purification methods are simplified and made cheaper, they can certainly live up to the theoretical claims of helping improve material qualities.

2.3.5 Similarities to GN

At this point, maybe we can introduce the idea of using GN as a low-cost substitute for CNTs of both types. Consider these points:

1. For catalysis and other charge-related applications, reactive edges of the GN behave as defective walls of CNTs, but with much higher surface area, and more importantly, this surface area can easily be quantified in GN by the BET test, while in CNTs, the surface area will vary based on the oxidation procedure and resulting defects, which could vary every time.
2. Heat treatment of platelet GN produces GN with closed ends, forming a "flattened" form of MWCNTs see (Section 2.5).
3. Cost of purification is lower, since by nature, graphitic material forms in platelet forms, i.e., CG form. The impurities, such as amorphous carbon deposits on the catalyst surface can be largely avoided by the use of proper reactor design (see chapter 4).
4. The low cost of production of GN might make some applications quite feasible.
5. In conductivity-related applications, ribbon type GN would perform well enough for industrial grade materials.

2.4 GRAPHENE OXIDE

GO was first announced in1859 by British chemist B.C. Brodie [110]. He added potassium chlorate ($KClO_3$) to graphite slurry in fuming nitric acid (HNO_3). The resulting material was composed of carbon, hydrogen, and oxygen, resulting in an increase in the overall mass of the flake graphite [111]. At 220°C, the C:H:O composition of this material changed to 80.13:0.58:19.29 ($C_{5.51}$:$H_{0.48}$:$O_{1.00}$) and lost carbonic acid.

Nearly 40 years later, L. Staudenmaier improved Brodie's $KClO_3$-fuming HNO_3 preparation by adding the chlorate multiple times over the course of the reaction (adding concentrated sulfuric acid), rather than in a single addition as Brodie had done. This slight change resulted in an overall extent of oxidation similar to Brodie's multiple oxidation approach but performed more practically in a single reaction vessel [112].

Sixty years after Staudenmair, Hummers and Offeman developed an alternate oxidation method by reacting graphite with a mixture of potassium permanganate ($KMnO_4$) and concentrated sulfuric acid (H_2SO_4), again, achieving similar levels of oxidation [113]. *The reason for this historical review is to point out that though others have developed slightly modified versions, these three methods comprise the primary routes for forming GO, and little about them has changed.*

We have learnt from our graphite discussion that a variety of oxidation products will be formed if this route is used for creating defects on the basal planes of natural graphite or CNTs. Epoxides, hydroxyl, carboxyl, peroxides, quinones, anhydrides, lactones, and many other compounds are formed on the basal planes by the harsh treatment. Today, most research publications concerning synthesis of graphene use GO as either a starting material or as an intermediate step.

Flake graphite is typically used as the raw material, and the resulting products from oxidation expand the graphite planes. The graphite is then known as Expanded Graphite of (EG) as shown in (Fig. 2.8A). The EG turns hydrophilic due to the oxygen groups, and if hydrophobicity is required for non aqueous functionalization, the EG is then reduced by hydrazine or other reducing agents, and exfoliated by ultrasound or other forms of energy. Or it is used as is for non-aqueous functionalization. In both these cases, the graphite suddenly changes its christian name to "graphene" with the surname Oxide. These references will be discussed when I discuss applications. Again, similar to publications on graphene, many "facile", "low cost" and "better" routes have been claimed [114-122] but the harsh and expensive Hummer's method or a slightly modified version of it is still the oxidation method predominantly used for creating defects on the basal planes of natural graphite or the walls of

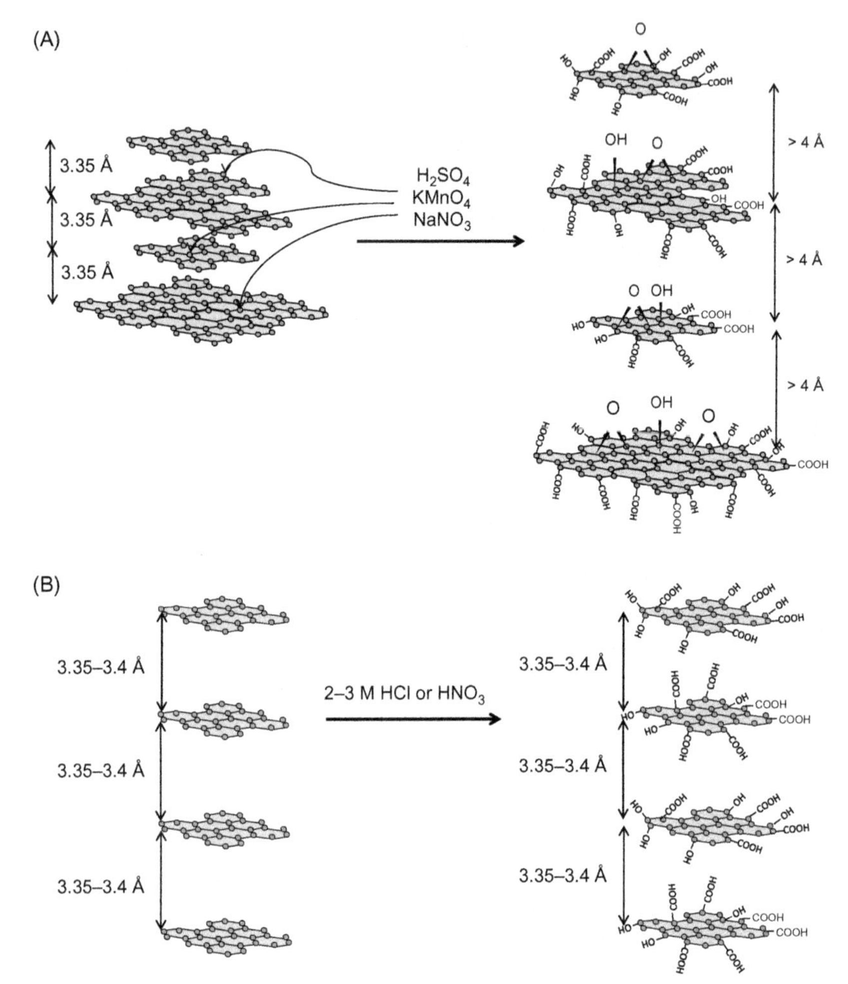

Figure 2.8 Graphical depiction of graphite oxidation and GN oxidation. (A) Graphite oxidation by the Hummer's method to EG. (B) GN oxidation by mild acids. Platelet type GN do not get attacked on the basal planes. *EG*, expanded graphite; *GN*, graphitic nanofiber.

CNTs to increase reactive surface area. Such harsh treatments are not needed when there is negligible basal plane related surface area, such as in GN. (Fig. 2.8B) illustrates the overwhelming dominance of edge site surface area in GN . The oxidized state of the edge sites in graphite or GN is almost exclusively made up of –COOH and –OH functional groups, and in the case of GN can be achieved with mild acid treatments.

In summary, for decades the procedure to make GO has hardly changed. There does not seem to be a substitute method on the horizon, either. The cost of making GO on a large scale is high and will be impossible to reduce if the Hummer's process is used, because most of the materials required are already commoditized chemicals.

2.5 GRAPHITIC NANOFIBERS

Carbon nanofibers (CNFs) are traditionally defined as cylindrical structures having an outer diameter below 1000 nm and an aspect ratio (the ratio between length and width) greater than 50. Carbon and graphite nanofibers are in their embryonic stage of commercialization. CNF can either have a hollow or solid section. CNFs with a hollow section have a microstructure similar in appearance to stacked cups or cones, and their inner walls are generally angled about 20 degrees to 30 degrees with respect to the fiber axis. The filaments have larger outer diameters compared to nanotubes (typically between 40 and 500 nm), and their length is commonly below 100 nM.

The tubular CNFs with smooth walls are generally useful for providing strength to materials due to the inherent strength characteristics achieved by the large aspect ratio (similar to the roots of a plant). The manufacturing process for such fibers usually involves carbonization of polymeric fibers followed by purification (baking) or other physical means to remove the soot and amorphous carbon formed on the exterior of the walls during synthesis.

The materials mostly discussed in this book can generally be described as GN. These fibers are of the stacked orientation, composed of graphite, the (002) planes of which are oriented in either the parallel (shell or ribbons), angled (fish bone), or perpendicular (platelets) to the axis of the fiber [123] as shown in Fig. 2.9. Thus, the surface area available from the edge region of these materials is theoretically infinite [124].

Characteristic of GN is a high concentration of terminal graphene edge sites on the external surface. The structure of GN is entirely composed of graphene basal planes. Nanotubes possess these planes in a rolled-up form, whereas GN possess them as flat planes with solely the edge sites available as active surface area. Three distinct GN types are shown in Fig. 2.9.

The fibers grow in both directions of the catalyst as shown in Fig. 2.9.

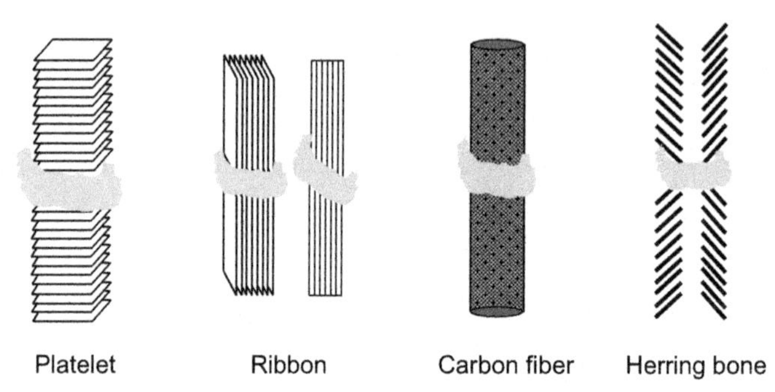

Figure 2.9 Types of GN—orientation of graphene planes. *GN*, graphitic nanofiber.

2.5.1 The ability of graphitic nanofibers to behave like SWNT and MWNT

Rotkin et al. [22] present a theory, describing the folding of a single graphene layer, which considers a free (not supported) monolayer that corresponds to suspended sheets of freshly cleaved graphite. While a free monolayer is needed to avoid consideration of the attractive van der Waals interaction between graphene sheets in the beginning of the folding process, the van der Waals forces cannot be completely neglected, because they define the shape of the final configuration. Benedict et al. discuss the collapse of a nanotube [29]. The interplay between van der Waals and elastic energies results in the formation of the "sleeve" structure along the edge with its radius of curvature being independent on the length of the edge. It depends neither on the sleeve orientation (chirality) nor on the presence of defects, postulating that the universal behavior can be predicted for the graphite from nanometer to millimeter scale. Folding of graphene is not an energetically allowed process because it starts with the rolling and increase of the elastic energy of the layer. The rolling of a graphene nano ribbon (GNR), resulting in closing of the dangling bonds at the opposite edges, is known to have a large energy barrier [125]. In the instance of the folding edges, the final state represents a lower energy owing to the van der Waals cohesion between the folded parts, but an external force is required to overcome the elastic energy barrier. After the curling layer has connected with the receiving layer, the (negative) van der Waals cohesion energy of the system grows proportionally to the contact area between the parts, thereby decreasing the total energy of the system. Then the layers stick together and the sleeve diameter decreases until the increasing elastic energy balances the van der Waals cohesion. The resulting structure has the diameter that is given by the ratio of the van der Waals–specific energy to the elastic constant. For the parameters taken by Rotkin et al. in their model, the optimum sleeve diameter was about 1.5 nm. A simulation was done to refine the continual result, which was confirmed qualitatively and was in good agreement with the experimental data on the size of monolayer nanoarches in graphite. However, the optimum radius is slightly larger, being about 1.8–2.5 nm, which is within a typical range of diameters for SWNTs. Thus, arches along graphite edges are predicted to have a curvature similar to SWNTs. The molecular dynamics (MD) simulation result for the zipping of the graphene edge is shown in Fig. 2.10. Similarities between a sleeve formed at the edge of graphite and an armchair SWCNT can be seen. The image in the background is a TEM image of a graphite polycrystal edge.

As explained by Rotkin et al., "dangling" edges in graphene planes will minimize energy by folding onto themselves given external energy such as thermal treatment. The sleeve opens and increases its diameter until the optimum geometry is reached. So, the flat closed double layer of graphene is recovered into the scrolled structure. GN of all types as well as flake graphite will thus exhibit this nucleating mechanism called "zipping" by Rotkin et al. It has been recorded that "zipping" for GN initiates at lower than graphitization temperatures, probably due to the synthesis process, and the purity of the graphitic materials, as compared to natural flake graphite. Complete edge closing was found when the GN were heat treated in argon at 1800°C for 1 hour (Fig. 2.11) [125].

Closing of the dangling bonds is energetically favored. Since the edges of the individual graphene sheets in GN are in registry, the "lip-lip" bond

Figure 2.10 MD simulation of GN edges zipping [23]. *MD*, molecular dynamics; GN, graphitic nanofiber.

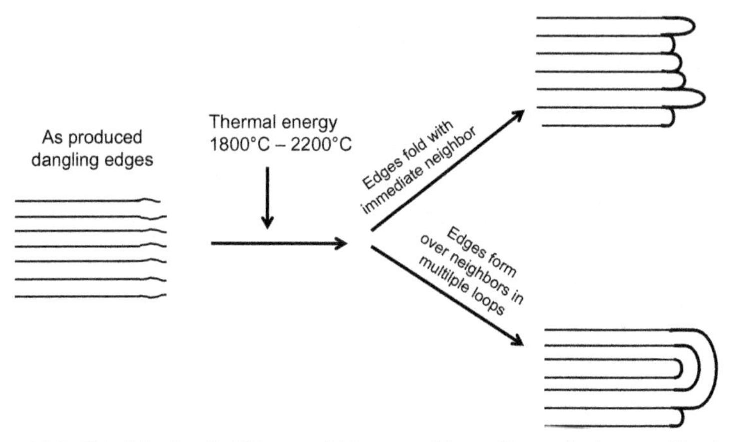

Figure 2.11 GN "zipping." *GN*, graphitic nanofiber. *From A. Lucas, Ph. Lambin, R.E. Smalley. On the energetics of tubular fullerenes. J. Phys. Chem. Solids 54 (1993) 587–593.*

formation is allowed [126], and a perfect nanotube-like structure grows [127], whereas in flake graphite and other less pristine graphitic structures, noncompatible edges form sleeves, which inevitably contain defects. In graphite, after the new bond formation, the edge is still far from equilibrium. The bond angles and lengths are not the same as they are for synthesized GN. That means the lattice is essentially deformed. To release this elastic energy, the edge has to pop up and form a wider sleeve, resulting in additional energy requirement because the van der Waals cohesion will be lost in the sleeve wall area. Finally, the sleeve arrives at equilibrium with the minimum energy, giving a mix of diameters and lengths for such structures.

The edges of platelet type GN form either single "scrolled" edges to form a "dumbbell" shaped graphene sheet, resembling a single-wall nanotube or multiple "lip-bonds" up to five to six layers of graphene sheets, resembling a "collapsed" multiwalled nanotube. Exfoliation would have to be figured out if they were to be made into SWCNTs. Heat treatment would automatically produce very CG structures with no defects. Naturally, the Platelet GN aspect ratios will be much smaller, and the applications competing with nanotubes would have to be ones that do not require high aspect ratios.

CHAPTER THREE

Graphitic Nanofibers—The Path to Manufacturing

3.1 CATALYSTS

The primary method for synthesis of carbonaceous nanomaterials that I will discuss in this book is what is known as Chemical Vapor Deposition (CVD). In the CVD process, (mostly) hydrocarbon gases are cracked to deposit elemental carbon on the catalytic surface. The catalysts can be in the form of immobile surfaces, such as wafers or patterned parts. They can also be in the form of granular particles. The granular particles are either in a packed bed configuration or in a fluidized bed. The fluidized bed configuration is the most practical configuration for a variety of reasons explained a bit later. Catalysts play a relatively important role in the manufacturing process. Many catalysts can deliver graphitic nanofibers (GN) with very high purity. In my personal experience, catalysts formulated and tested on a laboratory scale are typically not commercially viable to synthesize via the procedures outlined by the formulators. Process times, types of solvents, salts, temperatures and most importantly, ingredients with environmental concerns have to be modified to be feasible for large scale manufacturing. For our first licensed catalyst, we shopped around the world with catalyst manufacturers for contract manufacturing, but were told the procedures were unrealistic to replicate on an industrial scale. Industrial scale catalyst manufacturers start formulating with what is real and viable right from the start. Therefore, it is best to work with a manufacturer to formulate a commercial catalyst for the type of nanomaterial you intend to synthesize. Beware of extraordinary claims about the uniqueness of catalyst formulations. Most catalyst formulations are variations of previous work done by other scientists. More often than not, we can overcome the minor disadvantages of one catalyst over another by intelligent design of reactors, as well as being realistic in setting our commercial goals. Things don't have to be perfect to be useful and profitable. In our quest for mass production of GN, we discovered this fact at great expense.

Graphitic Nanofibers.
DOI: http://dx.doi.org/10.1016/B978-0-323-51104-9.00003-0

From this point on, I will discuss the rest of the topics with a specific focus on GN. The intent of the earlier discussions was to bring the key characteristics of other materials into focus. These are factors that show us the similarities between GN and the other nanomaterials, giving us the ability to extend the results and data of the published work (mostly with graphene in their titles) to GN.

Our team never looked at any catalyst that required gases other then methane or syngas ($CO + H_2$) for synthesizing GN. Our primary reasoning, as common sense would dictate, was the cost of production of GN. Methane, a component of natural gas, is available at a reasonable cost, but more importantly, it is a component of biogas sources such as landfill gas (LFG) or anaerobic digester gas (generated from oxygen-deprived digestion of carbohydrates, similar to the way our digestive system works.). The latter are derived from waste and available at a fraction of the cost of natural gas, not to mention they are immune to the energy price fluctuations, and maybe, just maybe, make you feel good about preserving earths resources. Similarly syngas can be produced by gasification of waste. Waste to energy is a rapidly growing industry worldwide. However, biogas, LFG and syngas need processing before use and result in end product being only slightly less expensive than natural gas given today's natural gas prices. We were fortunate to have had experience in gas purification and water purification to adopt this feed source. NG works nearly as well in terms of costs.

There are many compositions of catalysts that are in the public domain or formulated by scientists that are willing to share their knowledge, and help you guide your catalyst manufacturers at reasonable terms. These formulations can be used as a starting point to evolve into the product you want. Believe me you will have enough on your plate than having to worry about making catalysts with consistent properties. For our second round, we left it to the professionals. As a start, some cross-pollination with other industries that may use similar catalysts help you understand the basic mechanisms. Transition metals are the backbone of the catalysis field. They have been used extensively to produce soot-free products. Surely in that evolutionary process, the formulators must have learnt what "not to do" to avoid growth of carbonaceous fibers. We would simply start by doing the "not to do" and work toward an efficient catalyst.

3.1.1 Catalyst formulations

Carbon deposition on catalysts is very easy to do. Fine-tuning the chemistry and operating conditions for these deposits to take the shape and form is the task. Transition metals can catalyze the cracking reaction. Nickel, iron, copper, and zinc are the more feasible catalysts with good activities. Higher performance can sometimes be achieved by precious and other exotic metals, but by and large, precious metals are used only in reactions that are poorly catalyzed by the commodity materials. A variety of promoters may be added to the formulations. Promoters are surface additives that are not catalysts but help the selectivity or reactivity of reactions. A promoter can stabilize a valence state or a stoichiometry of reactants. A wide range of promoters are used in industry. The selection is based on the electronic structures of the reactants and catalyst. Potassium, titanium, molybdenum, copper, most lanthanide series, and their chloride salts are some that come to mind. There is a vast amount of information out there that can help one understand the functions of the different components, synthesis conditions, and operating conditions in the final process. Nevertheless, large-scale manufacturing remains an art more than science. Many formulations work well when you are making 1 g, but outside the laboratory, it is a whole different story. I list below some of the catalysts described in literature with regards to synthesizing GN. Reading between the lines, one can pretty much arrive at an intelligent starting point.

Ni-Mg catalysts have been widely used in reforming of light hydrocarbons and especially in the reaction of methane reforming with CO_2 to produce syngas ($CO + H_2$). Solid solution synthesis of NiO-MgO precursors for Ni-Mg catalysts has been reported [128]. Mainly they are nitrate salts oxidized to metal oxides or a slurry of metal oxides dried to produce the mixed oxides. Variations of these catalysts can be looked at, and have been utilized by many in the nanomaterials field, albeit sometimes reported as inventions.

Cu-Ni-Mg catalysts were used by Echegoyen et al [129] for methane decomposition and GN of 30 nM diameter were produced. The yields are mentioned in terms of hydrogen production (80 vol%). Ni-Cu-Mg catalysts lead to the formation of a high-order deposited carbon structurally close to perfect graphite, while catalysts in the absence of Cu lead to the formation of a low-ordered deposited carbon. A combination of the two above with only slight modifications, was licensed to us as an "invention".

Hulikova-Jurcakove et al. [130] outline tests for gravimetric capacitance of synthesized GN from methane and acetylene. They discuss the basic ingredients of the catalyst they used. When using methane as a feedstock, they produced GN with much higher graphitic nature, implying less surface defects. However the charge/discharge change in capacitance was higher with GN produced from acetylene that had higher surface defects on the graphene plane. Nevertheless the data from the methane-synthesized GN was far superior to the other carbon the two materials were compared with. To me, this is a perfect example of the "good enough" principle that guides the commercial world.

Ermakova et al. [131] described work done with Nickel including textural promoters such as oxides of Si, Al, Mg, Ti, and Zr. They reported high yields of nanofibers (>350 g/g of nickel) specifically with SiO_2 but medium yields with MgO and ZrO_2 textural promoters. The effect of various catalyst preparation steps is discussed in detail. Slight extrapolation of operating conditions such as increase in temperature to 650–700°C shows highly graphitic GN can be produced using some of the parameters discussed in their publication.

Reshetenko et al. [132] discuss the use of Cu-Ni-Al_2O_3 catalysts used for GN production. Two types of catalytic particles and correspondingly two types of GN were observed for the formulation 82Ni-8Cu-Al_2O_3 sample:

1. *Pear-shaped particles 40–50 nm in size.* These particles are reported by the authors to be typical for nickel–aluminum catalysts. The structure and morphology of GN shows that one carbon filament grows from one catalytic particle, and the graphite planes in filaments are arranged as coaxial cones. However the angle between graphite planes and the filament axis varied from 45° to 75°. This variation is not significant in the applications I will outline. The average diameter of the filament and the catalyst particle were identical (40–50 nm) as shown in Fig. 3.1A. The portion of the given particles is about 65% of the total range of the produced particle sizes.
2. *Quasioctahedral particles 70–100 nm in size.* In this case several filaments grew from one particle as shown in Fig. 3.1B. The diameter of the GN was less than the size of the "mother" particle. The graphene layers in the filaments are stacked perpendicular to their axis, what we term as platelet-type GN.

The authors report a yield of 525 $gGN/g_{catalyst}$ under a well-controlled environment with a maximum conversion of 40% and a 90 $L/h\text{-}g_{catalyst}$. All our economics and certainly our experience have been worked on ~50% of this yield and ~75% of the conversion rate. The replication of this data, even with our discount factors would be good enough for an economic justification, since the feed gas used was methane. The authors have tabulated further details about the catalyst compositions and the corresponding conversion rates and yields as shown in Table 3.1.

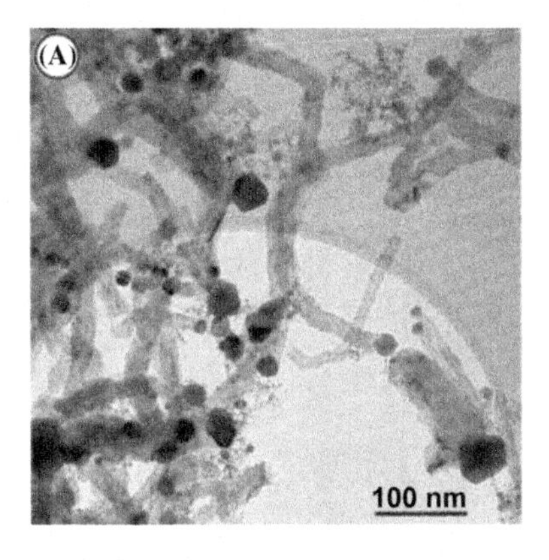

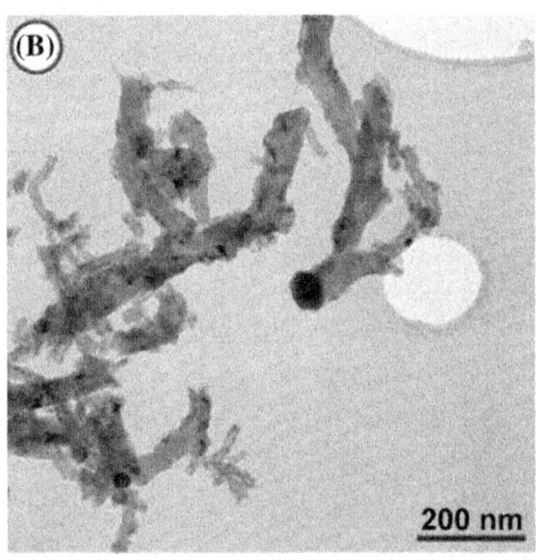

Figure 3.1 (A and B) Single and multiple fiber growth from catalyst particles. *From T.V. Reshetenko, L.B. Avdeeva, Z.R. Ismagilov, A.L. Chuvilin, V.A. Ushakov, Carbon capacious Ni-Cu-Al_2O_3 catalysts for high temperature methane decomposition. Appl. Catal. A 247 (2003) 51–63.*

Takenaka et al. [133] have also reported an impressive maximum yield of GN at 491 g/$g_{catalyst}$ with a Ni/SiO_2 formulation. With 40 wt% nickel, the crystallinity of the GN increased with temperature, but the yields dropped sharply after 10 hours online for temperatures above 550°C as shown in Fig. 3.2. The top portion of the graph shows the conversion rates for the first hour of operation (in fractions) and the bottom part shows the results of a longer run. We can see from Fig. 3.2 that while the low-temperature runs yielded more GN, the conversion rate at these temperatures was low from the start (~7.5–10%). At temperatures of ~600–650°C, we see a much higher conversion rate of 25% at the start, and the run ending at about 5 hours. Given the same selectivity, a higher conversion rate saves on feed gas, and 5 hours is not bad for a run in an

Table 3.1 Catalytic properties and yields of GN from coprecipitated Ni-Cu-Al_2O_3 catalysts

Sample	**625°C**			**650°C**			**675°C**		
	x **(%)**	***G*** **(g/$g_{catalyst}$)**	***t*** **(h)**	***x*** **(%)**	***G*** **(g/$g_{catalyst}$)**	***t*** **(h)**	***x*** **(%)**	***G*** **(g/$g_{catalyst}$)**	***t*** **(h)**
90Ni-Al_2O_3	31	22.4	2.0	34	11.2	2.5	7.0	8.0	5.0
82Ni-8Cu-Al_2O_3	22	515	61.5	29	169	15.5	35	150	9
75Ni-15Cu-Al_2O_3	18	525	54	23	400	37	27	404	27.5
65Ni-25Cu-Al_2O_3	15	291	35.5	20	309	32	26	293	20
55Ni-35Cu-Al_2O_3	13	205	31	17	207	26	20	222	19.5
45Ni-45Cu-Al_2O_3	10	118	23	15	115	13.5	17	126	12

Methane space velocity = 901 per $g_{catalyst}$-h, P_{CH4} = 1 bar.
Source: From T.V. Reshetenko, L.B. Avdeeva, Z.R. Ismagilov, A.L. Chuvilin, V.A. Ushakov, Carbon capacious Ni-Cu-Al_2O_3 catalysts for high temperature methane decomposition. Appl. Catal. A 247 (2003) 51–63.

industrial dual reactor environment. The yield is impressive, and again, with this conversion rate and even 50% of the yield they achieved, the financial numbers would make sense.

Li et al. [134] discuss formulations to demonstrate the importance of metal substrate interaction (MSI) on the catalyst deactivation rate during the methane decomposition to GN and the corresponding yield of the GN. The addition of BaO, La_2O_3, and ZrO_2 to the SiO_2 support of a 12 wt% Co/SiO_2 catalyst is shown to modify the reduction behavior of Co species and lead to changes in metal dispersion. The rate of catalyst deactivation during methane decomposition ($CH_4 = C + 2H_2$) is shown to increase with increasing MSI. The increasing rate of deactivation correlates with an increasing amount of graphitic carbon versus metal carbide on the used catalyst. It is suggested that an increase in graphitic carbon is a consequence of a strong MSI that limits carbon removal from the metal surface after filament formation.

Ashok et al. [135] examined the influence of copper content in a Ni-Cu-SiO_2 catalyst and report the formation of GN as a result of methane decomposition. At 650°C, the Ni-Cu-SiO_2 (60:25:15) composition displayed the highest activity with an initial conversion of 53%, a GN yield of 801 gC/(g_{Ni}), with a total run time of 1800 min. The GN

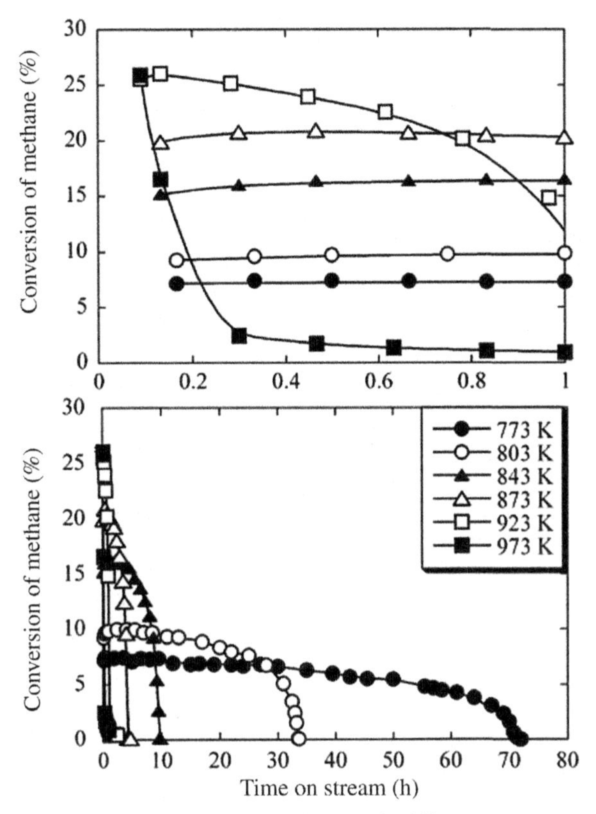

Figure 3.2 Conversion rates and time online with different reaction temperatures. *From S. Takenaka, S. Kobayashi, H. Ogihara, K. Otsuka, Ni/SiO$_2$ catalyst effective for methane decomposition into hydrogen and carbon nanofibers. J. Catalysis 217 (2003) 79–87.*

yield increased with an increase in the Cu content up to a molar composition of Ni:Cu:Si of 60:25:15.

3.1.2 Summary

The discussion above is very brief for the vast catalysis domain in any synthesis effort. But it should be enough to help connect the dots for those interested. The crux of the matter is that there are multitudinous avenues to end up with the catalyst most suited for GN synthesis, especially keeping the industrial needs in mind. Literature gives enough details to give anybody a running start to sit with their catalyst manufacturer and get to the required catalyst within a short period of time. Out of necessity, we changed from a licensed catalyst to our own when the licensor did not renew our license, and learnt all that I have explained in this section, with the pleasant conclusion of significantly reduced catalyst costs.

3.2 REACTION MECHANISMS

The specific properties of any type of carbon nanofibers (CNF) originate from their structure and dimensions. Generally the structure of GN is controlled by the nature of the catalysts used and the interaction between the catalysts and carbon sources under working conditions [136–138]. Therefore, in my opinion, some understanding of these factors is important to interpret empirical data and assist in the modifications of design parameters for your final equipment design.

The mechanism of carbon filament (fiber) formation over catalysts has been studied for many years [6,139–144]. However the work was originally done for reasons other than producing nanomaterials. In steam reforming and methanation processes, deactivation of catalysts or rupture of the reactor walls were found to be caused by the formation and/or deposition of filamentous carbons [139,145]. Since the 1990s CNFs have received new attention as potential nanomaterials. It has thus become important to devise means to selectively synthesize CNFs in as high yields as possible, instead of suppressing their formation.

3.2.1 Intermediate structures as building blocks

Before taking final form, it is postulated that the GN may go through an intermediate structure. Yoon et al. [146] examined structures of three typical types of GN, such as catalytically grown platelets, herringbone (also known as fish bone), and tubular types. Scanning electron microscope (SEM), high-resolution transmission electron microscope (HR-TEM), and scanning tunneling microscope (STM) were used for high-resolution scanning. A SEM is a microscope a beam of electrons scan the surface of materials. The reflection of these electrons form an image.

A HR-TEM creates images of the atomic structure of the material. It is one of the modes of an imaging microscope called Transmission Electron Microscope (TEM). In TEM, an electron beam is sent *through* an ultra-thin sample. The sample interacts with the electrons, and an image is formed. The image is then magnified and focused on an imaging device.

A Scanning Tunneling Microscope (STM) is a surface level atomic imaging tool. The STM is based on quantum mechanics principles, specifically the concept of tunneling. A conducting tip is placed extremely close to the sample surface. A voltage is applied between the two, and electrons, behaving like waves, can tunnel through the vacuum between them. The *current* generated is a function of tip position, applied voltage, and the local density states (LDOS) of the sample. The information from this current as the tip scans the surface is usually displayed in image form. Meso-dimensional clusters of the hexagonal graphene, such as carbon nano rod (CNR) and carbon nano plate (CNP), were found to be building blocks within the three structures.

In their study, highly graphitic GN were synthesized from CO/H_2 using Fe catalysts. Herringbone structures with lower graphitic nature were produced from C_2H_4/H_2 mixtures on Cu/Ni catalysts and graphitic tubular structures from CO/H_2 mixtures were synthesized from Fe-Ni catalysts. The SEM and TEM images are shown in Fig. 3.3.

The STM images of Figs. 3.4A and B show platelet GN as-prepared. Fig. 3.4C confirms these GN consist of a number of small units. After graphitization in argon at 2800°C for 10 min, the platelet GN were easily observed. Figs. 3.4D and E show a number of rod-shaped subunits (CNR), about 20 nM in length and 3 nM wide, which were packed perpendicular to the axis and formed a polygonal pillar. The ends of CNR had dome-like caps (Fig. 3.4E) Such caps corresponded exactly to loop-shaped ends, which form by the heat treatment of platelet GN, as seen in Fig. 3.4F. Fig. 3.4E also shows the hexagonal lattice of carbon atoms on the surface of an individual CNR. Fig. 3.4G shows uniform alignment of graphenes planes in the as-prepared platelet GN like that of the graphite, whereas the biased edges of its planes as indicated by dashed circles suggest that the concentric loop-ends as shown in Fig. 3.4F are not a unique product of graphitization, but a building block of GN. When GN were ball-milled in ethanol at room temperature, some rod units separated from the GN (Fig. 3.4H).

They also observed nanosized platelet units which they termed CNP as a building block in the same sample of the platelet GN, as illustrated with a model of hexagon-type plate stacking in Figs. 3.5A. CNP provided the same (002) lattice fringe pattern as that of GN or graphite from HR-TEM. The STM image in Fig. 3.5B and C shows the as-prepared platelet GN consisting of mainly CNP or its mixture with CNR. After heat treatment, the plate units became more distinct in their shape as

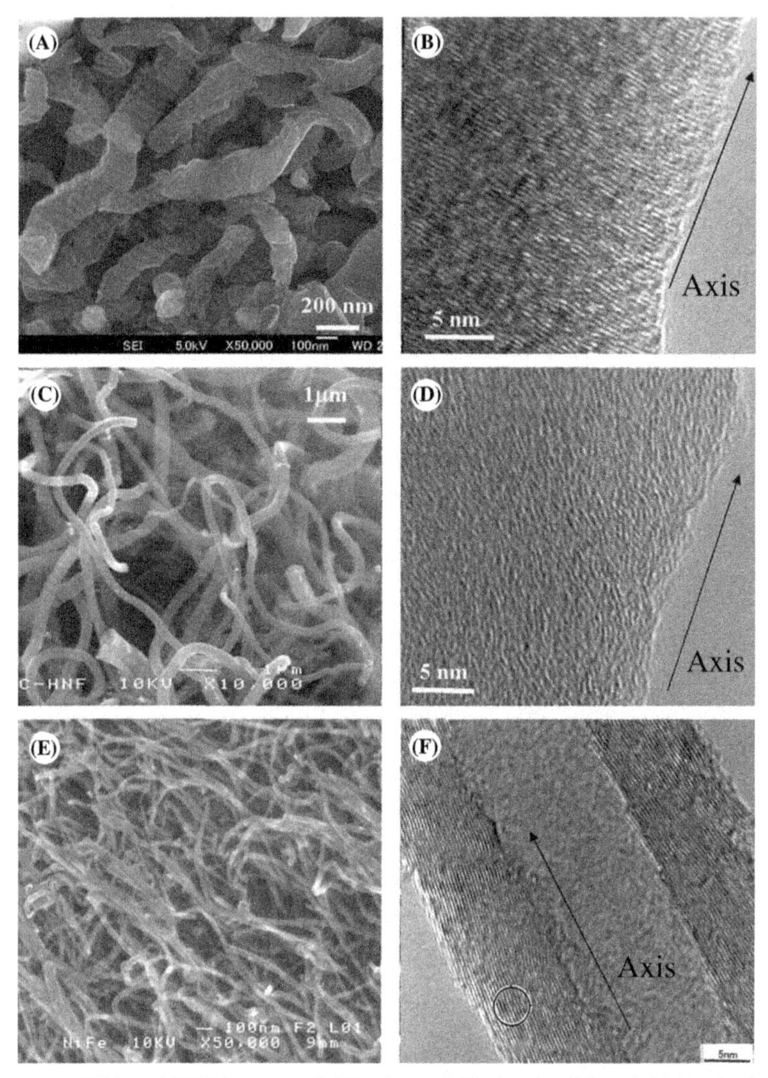

Figure 3.3 SEM and TEM images of GNs: (A and B) platelet, (C and D) Herringbone, (E and F) tubular [146]. *SEM*, scanning electron microscope; *TEM*, transmission electron microscope; *GNs*, graphitic nanofibers.

shown in Figs. 3.5D–I. Fig. 3.5D shows independent stacking units in the fiber, each of which probably consists of several graphene layers as shown in Fig. 3.5F. Figs. 3.5F and G suggest similar hexagon arrangements of elemental carbon in the A and B faces of a CNP to both equivalent basal planes of a typical graphite crystal. The STM image of

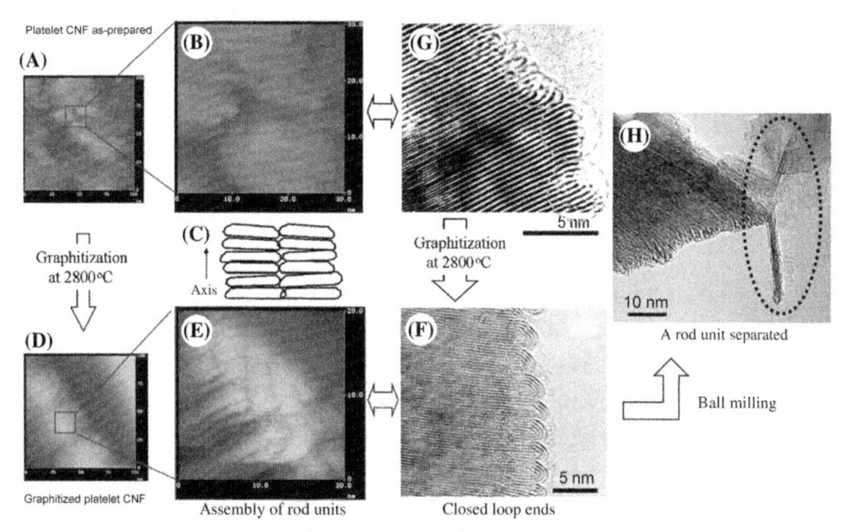

Figure 3.4 Rod-type unit as a constructive unit of platelet GNs: STM pictures of platelet GNs as-prepared (A and B), a schematic model of rod-type unit stacking (C), STM pictures of heat-treated GNs (D and E), HR-TEM picture of heat-treated GNs (F) and platelet GNs as-prepared (G), and HR-TEM picture of separated rod-type unit (H). *HR-TEM*, high-resolution transmission electron microscope; *STM*, scanning tunneling microscope; *GNs*, graphitic nanofibers.

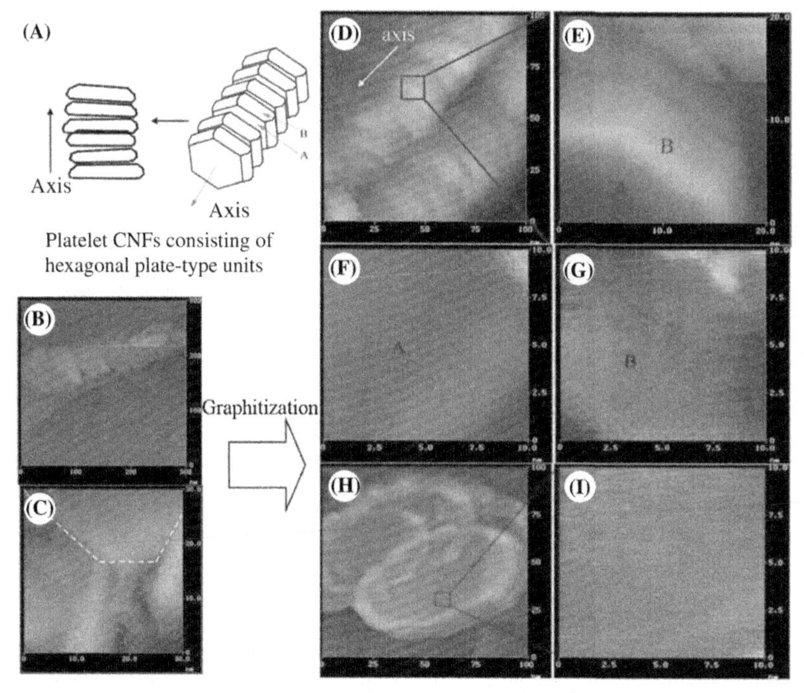

Figure 3.5 Plate-type unit as a constructive unit of platelet GNs: a schematic model of plate-type unit stacking (A), STM pictures of platelet GNs as-prepared (B and C), and graphitized platelet GNs (D–I). *STM*, scanning tunneling microscope; *GNs*, graphitic nanofibers.

Fig. 3.5H shows the obvious presence of transverse shaped polygonal plate units and of the surface of carbon basal planes (Fig. 3.5I). Based on these findings, the authors present a three-dimensional conceptual model. They hypothesize that the cross section of rods should be rectangular or hexagonal.

The authors propose that the formation and excretion of such a carbon cluster may not occur continuously but at intervals. The excreted cluster arranges as a stable form of carbon, assembling with others to produce a particular type of CNF. A catalytic framework within the metal particle may determine the dimension and morphology of CNP or CNR. This sequence would yield structural units such as CNR or CNP, subsequently forming GN.

3.2.2 Models of reaction mechanisms

Other typical models for the formation of GN involve the steps of surface adsorption/decomposition of hydrocarbons on the free surface of a catalyst particle, diffusion of carbon formed into the catalyst particle, and precipitation as graphitic layers on another surface of the catalyst [147].

- Baker et al. [148] correlated Ni, Fe, Co, and Cr with the activation energy of carbon diffusion in the metal and suggested that the diffusion of carbon through the metal particles is rate determining for filament growth. Moreover, they suggest that the driving force for this diffusion is a temperature differential created by the heat generated of the exothermic catalytic reaction supplying the surface carbon atoms and the heat absorbed by the endothermic reaction on the other face of the catalyst particle.
- But it has subsequently been pointed out [149] that filament growth was also observed in cases where the surface reaction was endothermic. Therefore, a carbon concentration gradient was suggested as the driving force instead. The gradient would occur due to a difference in carbon activities at the filament/metal particle interface and at the part of the metal particle surface where the decomposition takes place. This suggestion was supported by reports in the literature that the carbon activity in a metal depends on the composition of the gas phase adjacent to the metal surface.
- Boudart and Holstein [150] also demonstrated with calculations that the temperature difference between the exothermic and endothermic faces is negligible (less than 0.1 K). In the calculation it was assumed that the metal particle diameter is equal to (or less than) 100 nM,

which is an order of magnitude smaller than the particle size used in the CVD of GN. But in the case of GN synthesis via fluidized CVD, the fluidization of the particle exposes it to the same temperature on both sides. Furthermore, we see nanofiber growth on both sides of the catalyst. So, it would be reasonable to conclude that for most catalysts used in fluidized bed CVD synthesis of GN, the metal particle is in good thermal contact with the environment on both faces.

- Carbidea as intermediates. Geus and collaborators [151–153] have suggested what is known as the carbide formation mechanism. They reported on extensive studies of carbon filament formation from decomposition of CH_4 and CO over Ni/SiO_2 and Fe/SiO_2 catalysts by gas-phase analysis. They concluded from results of equilibrium studies and from estimates of the surface and defect energies of the filaments that the energy of filament formation or temperature gradient could not be responsible for the observed equilibria. They suggested that this conclusion together with the results of the measurements supported the idea of intermediate unstable carbide formation (carbide equilibrium model), which would determine the deviation from a graphite equilibrium.
- Snoeck et al. [154] proposed that the difference in the diffusional path length causes rapid nucleation and excretion of carbon layers near the gas/metal interface. The path length of carbon diffusion through the metal particle appears to depend on the nature of the metal particle, catalyst particle size, shape, and temperature. Further they suggest that, the interaction between metal and carbon sources at the gas/metal interface may affect the adsorption and decomposition of carbon sources, governing the nucleation rate.
- Helvig [155] and Puretzky et al. [136] used time-resolved, in situ HR-TEM and reported on the formation of fibers from methane decomposition by nickel nano crystals supported on Al_2O_3. The nanofibers were found to develop through a reaction induced spontaneous reshaping of the nickel nanocrystals into crystalline nickel nanoclusters. The alignment of the resulting fibers was dependent on the size of the clusters. The elongation of the Ni particles correlated with the formation of more graphene sheets at the graphene–Ni interface with their basal (002) planes oriented parallel to the Ni surface, offering the conclusion that the process must have involved *transport of C atoms* toward and Ni atoms away from the graphene–Ni interface. The nucleation and growth of the fibers involved a

continuously changing formation and restructuring of monoatomic step edges at the nickel surface with the nickel assuming elongated and pear-like shapes and returning back to near spherical shape in less than 0.5 seconds. Calculations with density functional theory (DFT) which I explained earlier, a computational quantum mechanical modeling method to determine the electronic structure in the ground state of a multi component system, confirmed that the growth mechanism involved the surface diffusion of carbon and nickel, but the metallic *step* edges acted as spatiotemporal dynamic growth sites. A graphene over-layer would change the adsorption energy of C and Ni adatoms on the Ni (111) surface. Nickel adatoms bound more strongly at the interface than on the free Ni (111) surface. Adsorbed C atoms could induce Ni step edges because the C binding energy to the Ni step is larger than the energy required for step formation. C adatoms at the interface would be destabilized increasing the barrier for carbon diffusion from the free surface to the interface. They conclude that surface transport of C atoms is the rate-limiting step for the nanofiber growth.

Though the initially described bulk carbon diffusion cannot be discounted yet, these results suggest that it is possible to transport C along the graphene–Ni interface. Step edges act as growth centers for graphene growth. Catalytic reaction designs usually assume a fixed number of stationary sites; so this finding of a spatiotemporal continuously changing growth center could give some insight on the yet so nebulous atomic level activity.

In my humble opinion, if we consider both mechanisms work together, we could possibly modify the Carbide Equilibrium Model slightly to conclude that it could be possible that the bulk diffusion of carbon through the catalyst experiences competion from the surface transport of C toward Ni edges. This combination could be constantly changing concentration of the carbide layer, leading to a small concentration polarization somewhere between the two metal faces that is overcome after more methane is decomposed on the surface, and more C is available. This small interval could explain the findings of Yoon et al. [146] that the fibers form in pulses.

Finally, Ammendola et al. [156] report on some practical factors such as resistance to attrition and pore occlusion. They tested what they term as high gas velocities up to 11.6 cm/s and found very little attrition. Judging by the van der Waal force strength, this result should not be surprising, since high Reynold's numbers would be difficult to achieve with moving particles of low density, low velocities, and short contact time, considering the vertical height of the reactors typically used for fluidized bed synthesis. As discussed later in the results of our work, the particles must be in a controlled turbulent mode for a longer period of time than simply the transit time through the reactor.

These critical factors are involved in the formation rate and type of CN precursors, and should be part of the thought process of design engineers in formulating catalysts as well as mass production reactors.

3.2.3 Summary

1. Diffusion through the catalyst particle is the most accepted path for GN synthesis.
2. Another widely accepted concept is the Carbide Equilibrium Theory, which predicts the formation of metal carbide films as an intermediate before diffusion into the catalyst surface.
3. Physical observations of intermediate structures that eventually can become GN, theorizes that the nature of GN formation is not consistent, and occurs at intervals.
4. A combination of **1** and **2** could explain the visual observations of **3**, and should be considered in the synthesis design of GN.

3.3 GROWTH RATES OF GNs

Knowledge of the growth rates with fairly close approximations is one of the factors high on the wish list among the important aspects of designing reactors. Mostly, this is accomplished by pilot studies and empirical data. However an intelligent starting point during the pilots never hurt anyone. In 2005, with limited knowledge and experience, our group needed some basic cause and effect knowledge for growth rates. We took guidance from engineering handbooks and literature but the task was unprecedented so we had to read between the lines on many fronts. We looked at CVD processes across different industries. Long story short, even then, we lost our catalyst from a vertical furnace for several of the initial runs.

One of the more comprehensive work I have read in this field was from Ilia et al [136]. They describe time-resolved reflectivity (TRR) of Vertically Aligned Nano Tube growth kinetics. Simply put, TRR is a way to study the change in properties (hence composition, phase) of a material over short periods of time (10^{-16} seconds) by spectroscopic methods using pulsed lasers. The paper elucidates the complex mechanisms and factors affecting the rate equation with direct measurements and a derivation of the resulting differential equation, finally leading to a simplified form that I could derive useful information from. Growth rates and terminal lengths were measured in situ. They considered the following factors for derivation and solving the rate equation:

- The number of carbon atoms initially formed by the cracking of the hydrocarbon gas (in this case C_2H_2).
- Of that number, the number of carbon atoms that formed a carbonaceous layer on the catalyst surface and did not diffuse into the catalyst particle and the inactive surface atoms of the catalyst on the surface itself.
- The number of carbon atoms that did diffuse into the catalyst particle.
- The number of carbon atoms that would precipitate as nanotubes by using the surface density of a monolayer of graphene and extrapolating to MWCNT.
- The activation barriers for sticking and catalytic decomposition of feed as well as their gas phase product.
- The flux of the hydrocarbon gas.
- Nanoparticle radius.
- Partial densities of feed and it's pyrolysis product that stays in the gas phase.
- Gas temperature.
- Mass of feed gas and that of the pyrolysis product.

A simplified model for an analytical solution of the rate equation is then presented by the authors for determining the parameters that control the growth kinetics and termination, leading to knowledge of rate constants. The simplified analytical solution for temperatures $<700°C$ and ignoring the effect of catalyst inactivity due to carbon deposition and a boundary condition requiring no conversion of feed to gas phase pyrolysis products, the rate relationships determined were:

The kinetics have the characteristic times, $\tau t = 1/kt$ and $\tau sb = 1/ksb$

Where:

kt = The characteristic rate constant for the bulk diffusion to the growth edge of a carbon nanotube.

ksb = surface–bulk penetration constant.

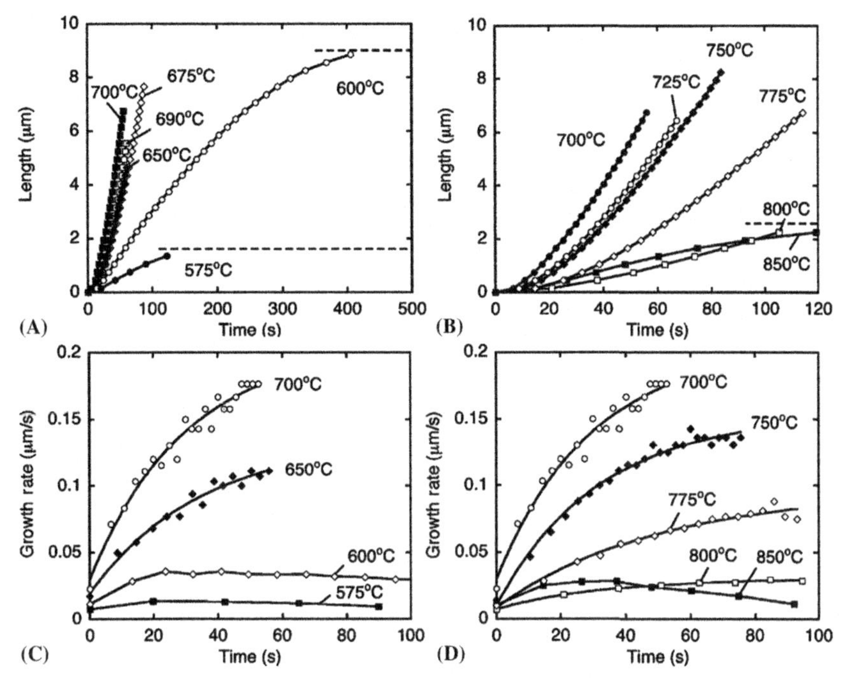

Figure 3.6 Growth rates versus time and temperature. Length of VANT (vertically aligned nanotubes) as derived from the experimental Fabry–Perot fringes for two sets of the growth temperatures: (A) 575–700°C and (B) 700–850°C. The dotted horizontal lines show the maximum achievable length of VANT restricted due to the termination of the growth: 1.6 μm at 575°C, growth time 518 s; 9.0 μm at 600°C, growth time 518 s; and 2.6 μm at 850°C, growth time 400 s. (C and D) Growth rates of VANT for the two sets of the growth temperatures calculated as a derivative from the curves shown in (A) and (B). *GN*, graphitic nanofiber. *From A.A. Puretzky, G.S. Jesse, I.N. Ivanov, G. Eres, In situ measurements and modeling of carbon nanotube array growth kinetics during chemical vapor deposition. Appl. Phys. A 81 (2005) 223–240.*

The termination time and length are shown to be defined as the ratio of the rate constants of bulk surface diffusion and carbonaceous deposit formation.

The authors also share growth rates for two different temperatures.

Their work on growth rates versus time and temperature is shown in Fig. 3.6.

We can deduce from the graph that at higher temperatures, it takes more time to reach the maximum growth *rate*. At 575°–600°C it takes less than 20 seconds to reach the maximum growth rate (C). At

650°C–700°C, this time increases to ~50 seconds. The published data on CVD reactor design, kinetics, and relationships between variables gave us a good start, but, as we shall note again in the manufacturing section, the best way to design reactors with the limited amount of useful theoretical information available is to collect empirical data with pilot studies. There is no way around empirical data.

CHAPTER FOUR

Manufacturing

4.1 METHANE AS A FEED SOURCE

I mentioned earlier that methane is the most economically viable carbon source for graphitic nanofiber (GN) production. Large-scale catalyst production, optimizing conversions and yields, purification toils, and capital unknowns provide enough challenges for cost control to even consider synthetic gases as feed. A good source of methane for small-scale GN manufacturing (~1–2 MT/day) is biogas. Biogas is generated by anaerobic digestion of organic waste. There are two major sources of biogas: landfill gas (LFG) and anaerobic digester (AD) gas. Both these biogas sources generate methane as a mixture with other gases. Methane content is generally 50–65% by volume, the rest consisting mostly of CO_2 and 3–4% of other gases. H_2S is always present, more in AD gas than in LFG. Therefore, unlike methane from natural gas (NG), biogas needs significant (but manageable) gas treatment to extract methane. The treatment costs nearly neutralize the advantage of the low cost of the raw gas (essentially free). The advantage lies in the predictability of the methane quality. By default, post purification, the hydrocarbon content consists of >99% methane. The synthesis process can tolerate some carbon dioxide and nitrogen. Elimination of carbon sources other than the one the catalyst is designed for is critical for producing a narrow specification product. When synthesizing nanomaterials, there is a delicate balance between the catalyst and the number of carbons in the feed gas molecule. As we have seen earlier, the reaction mechanism is sensitive to carbon concentrations at the gas–solid interface. Having more than a one carbon hydrocarbon in the feed gas when the catalyst was designed for methane or carbon monoxide (CO) can cause coking and deactivate the catalyst prematurely. NG can be <90% methane because it is sold on the basis of its heating value (pricing is in terms of million BTUs (MMBTU)). Compressed natural gas (CNG) is made from NG and is available in cylinders at ~3300 psig pressure. Natural gas (NG) can contain various percentages of many other hydrocarbons, the combined

Graphitic Nanofibers.
DOI: http://dx.doi.org/10.1016/B978-0-323-51104-9.00004-2

heating value of which are the only concern for combustion processes. Compressed natural gas (CNG) is made from natural gas, and is available in cylinders at ~3300-psig pressure. Short of an analysis, there is no particular standard or calculation of the methane content in these gases. When the methane content information is desired, one must ask for the Methane Number from their supplier. Again, the calculation to arrive at this number is not standardized, and each source has it's own formulation. Only thing for sure is that a methane number of 100 means essentially 100% methane. For synthesis, a methane carbon source must be virtually all methane to be able to make a consistent product. This severely affects the cost of the feed source for nanomaterial synthesis. By nature, when you purify LFG or AD gas, you produce high purity methane, but the purification cost is pallatable since the gas is virtually free. Possibly higher than the cost of purification, when you use natural gas with a low Methane Number, you will lose all the non-methane components during purification. The proportional cost increase and alternative uses of this waste gas must be planned for. For CNG, the cost issue is exacerbated by the energy required to compress and cool the gas. The cost of CNG could not be justified as feed gas for our analysis. If NG is chosen as the feed source, the production facility should ideally be located where the methane content in the supplied natural gas is ≥98% with the balance being devoid of hydrocarbons with more than 2 carbons. It is fairly easy to find natural gas with 90–93% methane. The higher the methane content, the lower the loss and impact on costs. In my cost analyses to follow in Chapter 5, I will assume a 90% methane content for calculating purification costs.

Between the biogas sources mentioned, our team chose LFG as our source of methane. LFG is generated as a result of anaerobic digestion of organic waste in closed landfills about 3 or so years after they are closed down. There are thousands of closed landfills in the United States generating methane in various amounts. With nearly 260 million tons of garbage being currently generated annually in the United States alone, this source of methane is not going away. LFG contains impurities that can be removed to a sufficient level for GN production. The details of our gas purification methods are beyond the scope of this book, but suffice it to say, purification can also be achieved with existing commercially available technologies. Once a closed landfill starts emitting biogas, the volumetric flow rate of the gas increases on a daily basis. When this rate first levels out, the methane fraction in the gas is high enough to justify the economics of power generation using internal combustion engines and turbines, or making CNG for vehicular use. In California, the LFG to CNG pathway has recently been qualified to be eligible for low carbon fuel subsidies (LCFS) which are similar in criteria and value to the Renewable Initiation Number (RIN) credits by the Federal Government. The increased financial incentives have generated a lot of activity in these fields.

At a later stage of a closed landfill's life, the methane fraction drops in the LFG with respect to the carrier gas and CO_2 fraction. The gas then has no commercially viable use at present. Due to the high Green House Gas value of methane (21 times more potent than CO_2), regulating authorities typically require the landfills to "flare" (combust) the methane and convert it to CO_2 and H_2O, the products of combustion, to reduce the GHG impact. Unfortunately, LFG usually exits the ground in short "puffs," and there is a short period between puffs that either has no gas flow or the returning carrier gas usually pumped into the landfill has no methane content. This issue necessitates the operation of a continuous pilot flame at the flare. Since most landfills are remotely located, this pilot flame is typically maintained by portable propane tanks at a significant expense. Fortunately, the methane concentration at this point is still significant on a mass basis to justify a GN plant for more than a decade at a proportionally sized landfill. The flare expense is eliminated and the (closed) landfill owner eliminates a cost center. There are thousands of closed landfills in the United States generating methane gas, making this a sustainable "green" source for feed with the additional benefit of low net cost.

Another advantage of using methane is the simultaneous generation of hydrogen gas as a byproduct, which can be a separate source of revenue, although we found the sale of hydrogen to be a very difficult task. Hydrogen could possibly have a use on site at some landfills, though. Typically the landfills have staged closures, and some parts of a landfill may still be emitting adequate methane fractions in their LFG, hence there is a possibility of either a power generation system or a CNG production facility at that location. If there are such scenarios, the operator of the power generation or CNG facility could benefit by injecting a small volume of hydrogen gas into their feed gas. Power generating engines produce more energy due to the higher flame speed of hydrogen. A small percentage of hydrogen can enhance NO_x reduction in SCR (selective catalytic reduction) units, which are required (at least in CA) to treat the exhaust gas from ICEs (internal combustion engines) [157]. At CNG production sites, injection of a small amount of hydrogen into

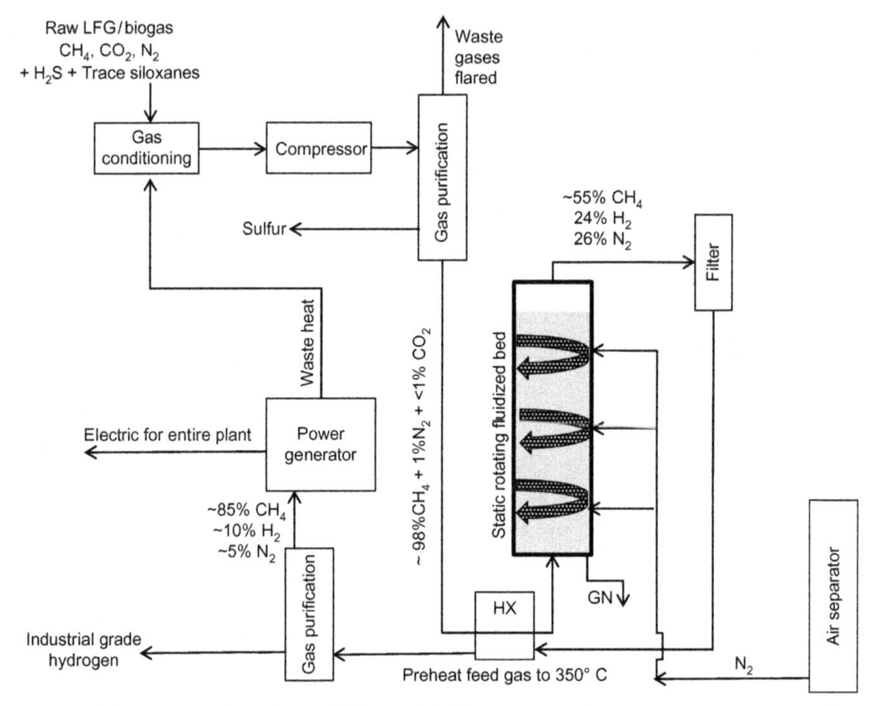

Figure 4.1 Process flow from LFG to GN. Pilot project by author's team at a San Diego County Landfill [158].

the methane would produce H-CNG. H-CNG engines have exhibited impressive reductions in regulated pollutants such as oxides of nitrogen (NO_x), non-methane hydrocarbons (NMHC), and carbon monoxide (CO). Reductions of 50%, 58%, and 9% respectively have been achieved with no fuel consumption penalty relative to CNG [157]. A basic process sequence diagram for a pilot GN production plant that my team ran at a San Diego County Landfill is given in Fig. 4.1.

4.2 SYNGAS AS A FEED SOURCE

Another single carbon source is synthesis gas. Many of the manufacturing processes for GN can use synthesis gas for the production of CNT and GN. Synthesis gas can be produced relatively inexpensively by gasification of carbonaceous solids. Coal gasification processes to produce syngas have been widely used in the world for more than half a century. Waste biomass can also be gasified for syngas production. There is a worldwide drive to reduce the load on landfills. Municipal solid waste is gasified in many countries to generate power with the produced gas (syngas). There is a growing trend across all continents to install small-scale gasification systems to enable widespread use. Small volume generators of biomass (agricultural industry) can capitalize on their waste. In some states, such as California, agricultural waste as well as livestock waste disposal is a persistent problem. Smaller, distributed generation systems could provide multiple sites for the production of GN. Syngas generation production cost would be pretty close to GN production from LFG, with the possible benefit of having a negative cost, such as tipping fees from landfills, etc. (Fig. 4.2).

In either case the production cost from waste generated gases is fairly close to that of producing GN with high-purity NG. If significant volumes of C>2 hydrocarbon gases must be removed from the NG to meet Methane Numbers, the feed cost from NG becomes significantly higher than biogas or syngas.This is primarily due to the volumetric loss

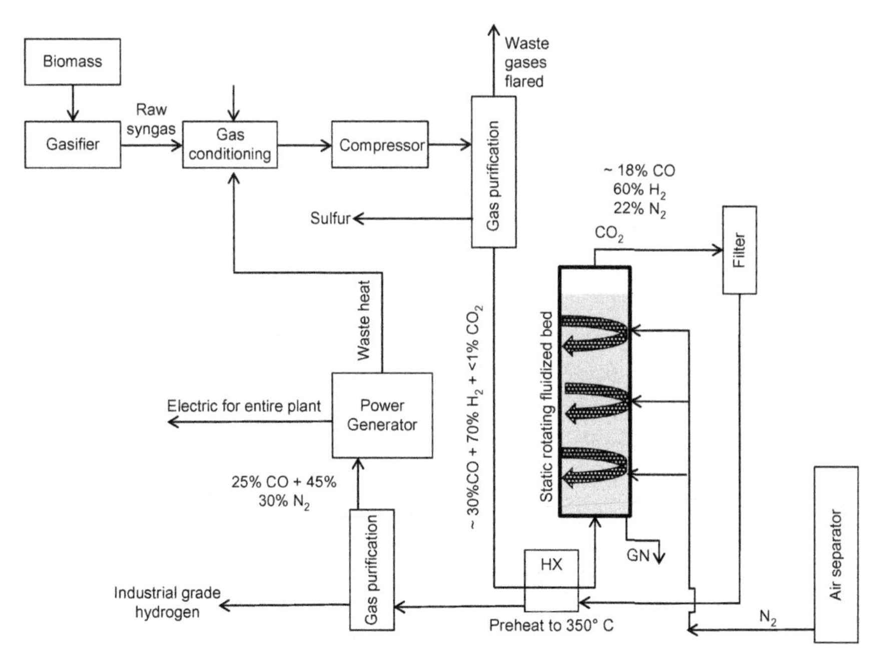

Figure 4.2 Conceptual process flow for manufacturing GN from syngas generated by waste biomass gasification.

resulting from the removal of these undesired hydrocarbons. I do not have data on such a purification process and the associated effciency losses which would be in addition to the loss of hydrocarbon gases. The overall advantage of using methane or syngas is that the power and thermal energy can be generated on site. Remember that overall conversion rates for GN and most other carbonaceous nanomaterials are ~35%. Besides, the unreacted gas is high-quality methane or CO with hydrogen. These gases may be used as fuel for power generation. The temperature controls required in synthesis reactors are best achieved with electrical heating anyway. Utilizing product gases from the reactor essentially eliminates additional process energy costs.

4.3 PRODUCTION

The original report by Ijima in 1991 [6] about highly graphitized carbon nanotube describes the Arc Discharge method at high temperature (>3500°C) and very short microsecond pulses. The small quantities produced and high-energy consumption were the drawbacks of this original process, though it was the process of choice for many years, especially by research organizations and scientists.

The most viable method for mass production is the fluidized bed chemical vapor deposition (FBCVD) process. The first fluidized bed gas generator was developed in the 1920s. An industrial scale reactor was first utilized for catalytic cracking of large, complex petroleum molecules to simpler compounds by what is now Exxon Mobil. This fluidized bed reactor (FBR) and the many to follow were developed for the oil and petrochemical industries.

An FBR can be used to carry out a variety of multiphase chemical reactions. In this type of reactor the reactant gas is passed through the catalyst particle bed at high enough velocities to suspend the catalyst and cause it to behave as though it were a fluid. The resulting fluidized bed imparts many important advantages to the FBR.

For the FBCVD reactor, the once-through vertical configurations are most common, and as such we will discuss the design of a couple of large-scale production reactors known in the public domain via patents or publications. My attempts to discuss manufacturing details with other overseas manufacturers that advertise low-cost nanomaterials were in vain.

I will first discuss some basic parameters briefly to lay the groundwork of the limitations and possible solutions to the synthesis of GN or any other nanocarbonaceous material in a vertical FBCVDR. Many factors would conceivably affect the overall success of the process. In the following sections I will outline the influence of:

1. Minimum fluidization velocity.
2. Catalyst size and effects of GN growth and agglomeration.
3. Agglomeration.
4. Reactor types.

Temperature and partial pressures are also important parameters but should generally be known from the nature of the catalyst and empirical lab data before a reactor is designed.

4.3.1 Minimum fluidization velocity

Many publications exist decribing computer models predicting the minimum velocity that is required for the fluidization of the bed [159–161].

Mostly, industry deploys vertical. The calculation for the minimum velocity for vertical FBR (where no deposition occurs) is straightforward. We first balance the gravitational force of the catalyst particle and account for the volume of the void space desired, correlated with the cross-sectional surface area of the reactor. The net force on the bed would be the pressure drop (ΔP). The final void volume determines the contact time, probability of reactant/catalyst contact, and more importantly, the height of the reactor to avoid catalyst loss from the top of the reactor. In most cases the catalyst particles are small, resulting in low gas velocities overall. Hence, we can neglect some theoretical components like frictional drag along the walls, etc.

Initial void volume prior to fluidization is the interstitial space between any set of solid particles. For perfectly spherical particles, a tetrahedral void is formed. Chemical engineering handbooks estimate this void volume to be 40–45%, but during my water treatment years, when we used uniform bead size ion exchange resins, the actual working void volumes were noticed to be ~35%.

The net catalyst volume is expressed as:

$$V_c = (V_x) \times C_{SA} \times H$$

where

V_c = total volume of catalyst particles in the reactor.

V_x = volume fraction of catalyst particle (100%-void volume %)/100.

C_{SA} = cross-sectional surface area of the reactor.

H = height of the unexpanded bed of catalyst.

Then the weight of the particle (downward gravitational force) can be calculated as:

$$W_p = V_x(\rho_c - \rho_f)(H)(C_{SA})(g)$$

where

$(\rho_c - \rho_f)$ = density difference calculated by bulk weight divided by the difference between bulk volume of catalyst less the pore volume within the catalyst particle.

g = gravity in appropriate units.

For minimum fluidization velocity, this force must be equal to the upward force necessary, which is the pressure drop across the cross-sectional surface area of the reactor. Equating the two:

$$\Delta P = V_x(\rho_c - \rho_f)(H)(g)$$

The typical particle sizes have a small Reynold's number ($Re \leq 10$). The Kozeny–Carman relationship (from *Chemical Engineering Handbooks*), which is used for viscous fluids, can be used to establish the initiation of fluidization:

$$V_f = \frac{(\rho_c - \rho_f)gD_p^2(1 - V_x)^3}{(150\mu)(V_x)}$$

Beyond fluidization, a standard FB reactor design would include the prevention of the catalyst particles being expelled from the reactor due to excessive gas velocities. This condition imposes a superficial velocity limit to always be lower than or equal to the settling velocity of the catalyst particles. Since we are dealing with small catalyst particles, Stokes' law can be used to calculate the settling velocity:

$$V_{set} = \frac{(\rho_c - \rho_f)gD_p^2}{18\,\mu}$$

The ratio can thus be derived as

$$\frac{(V_{set})}{(V_f)} = \frac{(25)(V_x)}{3(1 - V_x)^3}$$

For theoretical values of 0.40–0.45, this ratio ranges from 78 to 50. At 0.35, it can be as high as 125. As mentioned before, the correct bed expansion design is achieved with other parameters also in mind, which relate to the surface chemistry of the catalyst.

For FBCVD reactors, the equations get more complicated. To start with, we don't have a bed of independent particles to fluidize with a large initial void volume. On the contrary, we actually need to have controlled attrition in the bed, to scrub the formed fibers off the catalyst surface. In the case of GN growth, we have an additional force to overcome, between the GN attached to the catalyst particles as well as the floating ones. The dangling edges of the freshly formed graphite are very active and tend to agglomerate rapidly, to achieve the lowest energy state, leading to occlusion of the catalyst particle surface. Therefore these reactors must be designed not only beyond the minimum fluidization velocity, but also in the turbulent regime.

4.3.2 Catalyst particle characteristics

4.3.2.1 Size

In heterogeneous catalytic reactions the particle size of the catalyst plays an important role. The surface diffusion and surface movement of carbon atoms are affected by the surface area of the particle, while the diffusion of the carbon particle and transport of it to the excretion face is dependent on it's pore size and pore volume. In FBCVD reactions the gas velocity and the turbulence created by the velocity add another degree of freedom to the picture.

Let us examine what we desire. We would like maximum or slightly pulsed contact of the catalyst with the reactant gas. If the collision of the particles with the reactant gas is not continuous, perhaps a little more time is afforded to the diffusion phenomena discussed in Chapter 3. At the same time, we would like the pore size and volume of the catalyst to be large, again, to make the diffusion and carbon atom transport faster and easier, but there is an upper limit to that parameter. The pore size will have limitations on the wall thickness of the pores. Very large pores sizes can cause the particle to break from attrition. A turbulent environment with the appropriate volume ratio of reactant to carrier (inert) gas would then enhance the process to an upper limit. A turbulent environment with a small particle size would also provide better heat transfer. The above parameters combined with a smaller surface area would also enhance the deposition process, by reducing the probability of the incidence of amorphous carbon deposits on the surface.

Pell M. [162], and Ray et al. [163] pointed out that the catalyst particle size influences fluidization parameters including minimum fluidization velocity, the terminal velocity, solids expulsion rate, reaction efficiency, and also hydrodynamic behavior. Derivations and studies done on fluidized bed dynamic behaviors focused on the bubble state, which is reached just prior to the minimum fluidization velocity [164].

4.3.2.2 Dynamic changes in size and volume

In the synthesis of GN, various sized structures can form as a result of the instability of two-phase fluidized flow, especially when the reaction generates a gaseous by-product, as is the case with methane cracking for GN production. Two moles of hydrogen gas are produced for every mole of carbon diffused onto/into the catalyst particle. The particles invariably have a distribution of sizes that evolve naturally through attrition and varying fiber lengths. If at all possible, accounting for the particle size distribution (PSD) could help tremendously in predicting the rate of possible expulsion of particles from turbulent fluidized beds, cyclone efficiency, etc.

4.3.2.3 Fluid–particle drag force

We have already discussed the gravitational force exerted on the particles in the Section 4.3.1. Another force, fluid–particle drag can be critically important in modeling of reaction kinetics. A number of empirical constitutive models for the fluid–particle drag in homogeneous suspensions of uniformly sized spherical particles are available in the literature, albeit for bubble type fluidized beds. A practical difficulty comes in applications with changing particle shapes and sizes instead of the uniformly sized, spherical particles that are free of deposition during the reaction. Drag laws for homogeneous suspensions of particles having a distribution of sizes are described in detail by Van der Hoef et al. in literature [165].

4.3.2.4 Size distribution

Distribution of particle sizes for spatially and temporally changing structures such as GN is handled in one of the approaches of the Euler model, by what is called the Discrete Quadrature Method (of moment). Quadrature based methods are similar to computational fluid dynamics, but are better suited for simulating phases such as dispersed phases in a multiphase flow. The smallest "particle" entities which are tracked may be molecules of a single phase or granular "particles" such as our catalyst particles, with and without the GN growth. A sum of delta functions is placed at selected particle sizes (quadrature nodes), these quadrature nodes are allowed to change temporally and spatially to capture agglomeration and break-up, therefore changes in size by chemical reactions, etc. The number of different particle phases that are needed to capture the effect of PSD will clearly depend on the nature of the PSD, so growth rates and time functions become important in our design of the GN reactors [166].

A simpler simulation approach that may be applied to low volume fractions of catalyst particles (as in the case of GN synthesis), is the Parcel Approach where Newton's equations for the motion of particles are derived. The Euler equation can be utilized for the particle momentum (gas velocity and pressure differential). The Euler equations govern the conservation of mass, momentum and energy in flows within adiabatic and viscosity free regimes. At very low volume fractions where interparticle collisions are less probable, initially one can ignore collisions and employ a point-particle approximation. At volume fractions of interest to us, parcels of point particles are simulated in this approach. Here each test particle being tracked represents a large number of particles having the same characteristics as the test single point particle [167,168]. The approach needs modification when large catalyst volume fractions are involved. Our focus for GN would not require us to delve into this detail. Though still in a theoretical model stage, it will be a great tool in the future for estimating PSD. Therefore, we conclude:

1. It is much easier to handle PSD with single-sized particles with no change in size during operation than in multifluid models. Unfortunately, the sizes of the GN synthesis catalyst particles are constantly changing and the particle is constantly under a spatial and temporal momentum influence.
2. In dilute flows, boundary conditions are easily implemented.

The parcel-based approach requires use of software to solve. There is commercially available software, MFIX® and Fluent® to facilitate the calculations. Once again, these models are important in the determination of the influence of the drag forces on the changing volume during the growth of GN and in the determination of optimum catalyst volume fractions. No experimental data is available to my knowledge to validate these model predictions. For now, empirical data has to suffice for design.

4.3.2.5 Agglomeration

The interaction between graphitic structures due to dangling bonds is a significant factor in the synthesis of graphitic fibers via CVD. The forces cause agglomeration between fibers (lowering dangling bond energy) and create stearic hindrance, resulting in complete blockage of the reactant gas pathway to the surface of the catalyst particle. Or, they are passivated by the hydrogen gas. FBCVD processes make it possible to create turbulent regimes within the reactors for the fibers to have minimum collisions, while simultaneously dislodging the fibers from the catalyst particle. Furthermore, once they are dislodged, the fibers can easily be removed from the reactor as product due to their lower density compared to the catalyst particles.

Since the van der Wal force is an important factor in all the carbonaceous nanomaterials, albeit more so in CNTs, I am covering it in some detail. VdW force is a derivative of the ideal gas law, incorporating molecular size and additional intermolecular interaction strength in the equation.

$$(P + a/V^2)(V - b) = RT$$

where a is the strength of the intermolecular interaction, and b is the molecule size.

The forces consist of orientative, inductive and dispersive components. Dispersion is the dominating force in the attraction between graphene planes. Fritz London in 1927 proved that dispersion is a quantum mechanical phenomenon, with spontaneously induced dipole moments.

This characterization applies universally to polar as well as nonpolar molecules. For graphite, Wang et al. [169] determined experimental values for the exfoliation energy between two graphene layers in graphite. Interestingly, they report (though on a basal plane dimension) that the cleavage (exfoliation) energy is independent of the twist angle and the temperature. Their data was collected with twist angles $16° < \Phi < 54°$ and temperatures between 0°C and 200°C. They reported measured values of $0.37 \pm 0.02\ \mathrm{J/M^2}$. These experimental values can give insight for designing turbulent FBCVD reactors. For agglomeration between graphitic surfaces such as graphitic nanofibers (GN), a mechanical force combined with low collision probability could well define the flow regime for the reactant gas in our reactor.

Theoretically the continuum Lenard-Jones (LJ) model suggested by Girifalco [170] is usually used to evaluate the potential between two graphitic structures. The LJ potential for two carbon atoms in graphene–graphene structure is

$$\varphi C - C(r) = A\, r^{-6} + B\, r^{-12}$$

where r is a distance, A and B are the attractive and repulsive constants. The integration of the LJ potential for crossed graphitic CNT was performed by Zhbanov et al. [171] with an average carbon atom density

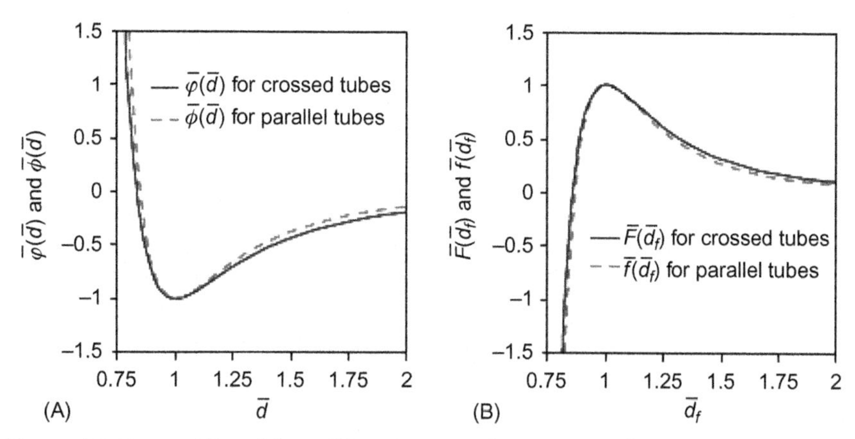

Figure 4.3 Energy (A) and force (B) and potential curves as a dimensionless distance parameter. *From A.I. Zhbanov, G. Pogorelov, Y.C. Chang, Van der Waals interraction between two crossed carbon nanotubes, Research Center for Applied Sciences, Taiwan.*

of $4/\sqrt{3}\ a^2$, a being the lattice constant for graphene hexagonal structure = 2.46 Å. They have presented uniform curves for potential and forces between two SWNT (see Fig. 4.3).

4.3.3 Reactor types

As I mentioned above, in a FBR the reactant gas is passed through the catalyst particle bed at high enough velocities to suspend the catalyst and cause it to behave as though it were a fluid.

Gas distribution: A porous plate designed to have uniform pressure drop (typically 0.5–1.5 psig) across the pores provides a uniform volumetric flow within the reactor, and is known as a distributor. The gas is forced through the distributor up through the catalyst bed until minimum fluidization velocity is reached and the catalyst bed enters the fluidized state. Depending on the reactor operating conditions and the properties of the catalyst, various flow regimes can now be observed in the reactor.

FBRs are relatively new tools for the deposition and phase change catalytic reactions in the chemical engineering world. Today FBRs are still used to produce gasoline and other fuels, along with many other chemicals. Primarily, FBRs are configured in three versions:

1. Once-through vertical tube reactors with periodic catalyst removal, and the unreacted gas being reused (after purification, if applicable) or, catalyst recycle with internal or external centrifugal separators.
2. Recirculating bed reactors.
3. Rotating FBRs.

4.3.3.1 Once-through reactors

Many industrially produced polymers require continuous operation, and the products are usually gaseous in nature. Such applications may use the reactors with internal and external solids (catalyst) separation mechanisms to allow the gas to exit from the top, while solids are either returned back to the reactor, or replaced with new catalyst. (see Fig. 4.4).

FBRs are also extensively used for gasification of biomass and coal to generate synthesis gas [172]. Hot sand is continuously circulated between two chambers for the endothermic process, while the main chamber is kept fluidized and biomass is fed in for gasification.

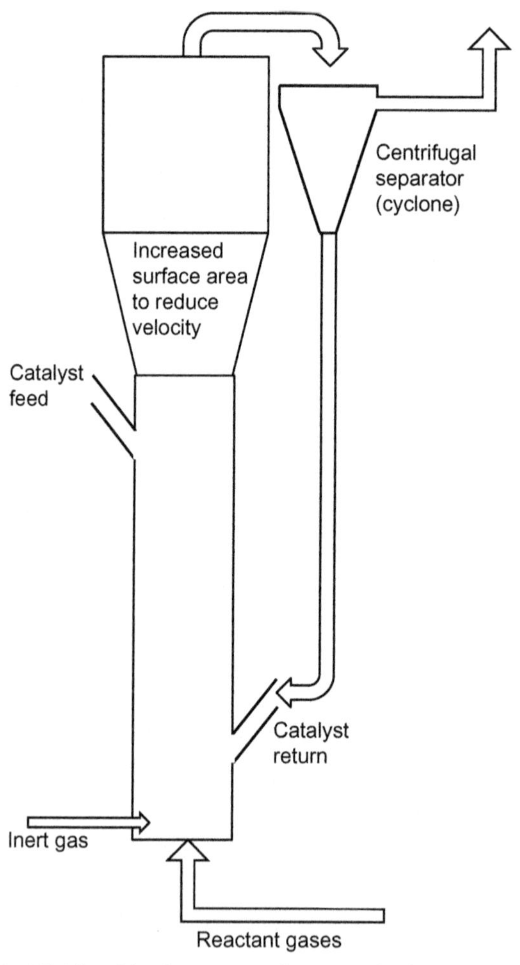

Figure 4.4 A typical fluidized bed reactor with external solids separator.

4.3.3.2 Recirculating FBRs

Recirculating FBRs continuously recirculate the solids or catalysts in a loop, without removing any solids (Fig. 4.5). When a slight change in equilibrium with an external chemical or material regenerates the solids media, this process can be very effective. Catalysts may get reduced during the process and simple aeration may be sufficient to regenerate them. In such reactors, air may be introduced prior to the catalyst coming in contact with the reactant gases for the reaction cycle. Another application may be in liquid–solid interface reaction applications such as ion exchange used in water treatment and other regenerable adsorbent processes. Developed originally by Irwin Higgins in the early 1950s for radioactive ionic separations, it has become popular for many ion exchange processes where ionic species need to be separated from (mostly) aqueous solutions which contain concentrations of the target

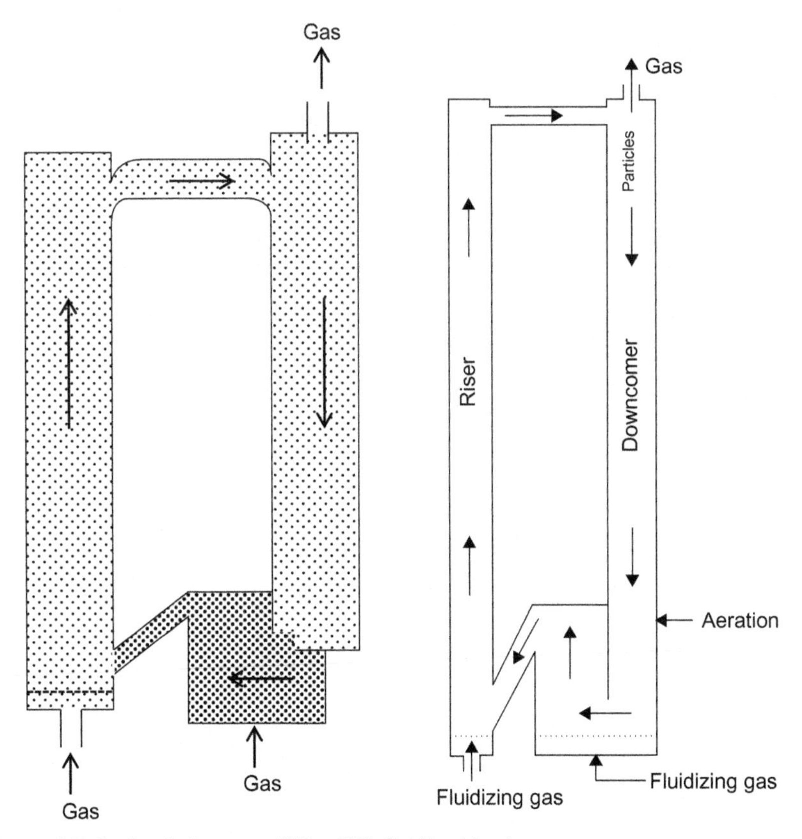

Figure 4.5 Recirculating type FBRs. FBR, fluidized bed reactor.

ion. The fluidized and recirculating ion exchange resin bed is regenerated at one point of the recirculation loop, rinsed in the immediate point following, and returned to the treatment cycle in the next node.

4.3.3.3 Rotating fluidized reactors

There are two main types of rotating fluidized reactors:

1. Rotating chamber.
2. Static chamber.

4.3.3.3.1 Rotating chamber type reactors

The reactors discussed so far may be called gravitational in nature. Centrifugally driven fluidized beds or rotating fluidized beds (RFB) offer an alternative technology where the particles are fluidized by balancing the radially acting drag and centrifugal forces. The centrifugal force, unlike the gravitational force, can be adjusted by changing the rotational speed of the particle bed or by modifying the reactor diameter. As such, RFB can handle much higher fluidization gas velocities since higher drag forces are allowed. The rotational movement induces a centrifugal force that drives the particles toward the cylindrical wall of the fluidization chamber to form a fluidized annular region.

In the rotating chamber RFB, [173] a the centrifugal force is generated by rotating the reaction chamber around its axis while the gas enters radially through the cylindrical wall of the fluidization chamber, similar to a washing machine or clothes dryer (see Fig. 4.6). The unit can be

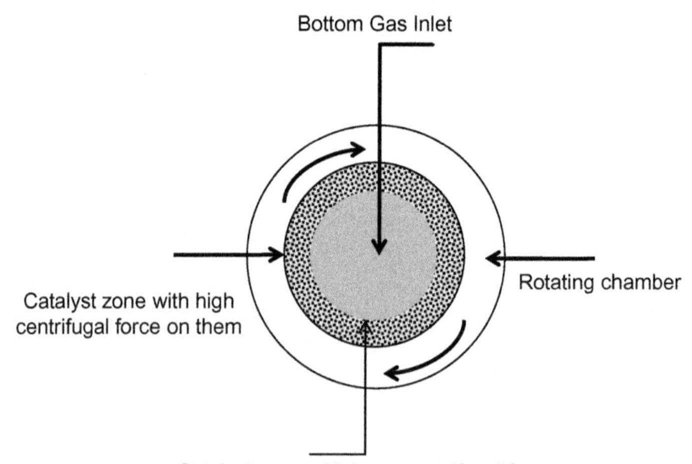

Figure 4.6 Rotating chamber FBCVD. *FBCVD*, fluidized bed chemical vapor deposition.

operated using either a horizontal or a vertical axis of rotation. Chen [173] showed that in an RFB, the particulate material is fluidized layer-by-layer, starting from the solid–gas freeboard, with increasing radius, as the gas velocity is increased.

4.3.3.3.2 Static chamber type reactors

A static, rather than rotating chamber FBR generates the centrifugal force by the tangential injection of gas through slots in the wall of the reactor [174–178]. The gas is forced (in a radial motion) to leave the fluidization chamber via a top aperture in the center of the reactor, commonly called a chimney. While the tangential gas velocity component is at the origin of the tangential fluidization of the particle bed and the centrifugal force, the radial gas velocity component induces a radial drag force causing radial fluidization of the particle bed. Despite higher fluidization gas velocities, better and adjustable gas–solid separation is achieved due to the adjustable centrifugal force [179]. The gas–solid contact time is much shorter, because the bed thickness is relatively small while the gas flow rate is high (see Fig. 4.7).

For deposition reactions such as uniform layers of materials on ceramics and aluminum as well as in the synthesis of nanostructures, FBCVDR technology has evolved very rapidly since the late 1990s. Minor variations of the process now can produce GN and CNT with very high consistency and desired forms. When designed properly, high-purity GN can be synthesized. Catalysts used in nanomaterials production via any CVD process

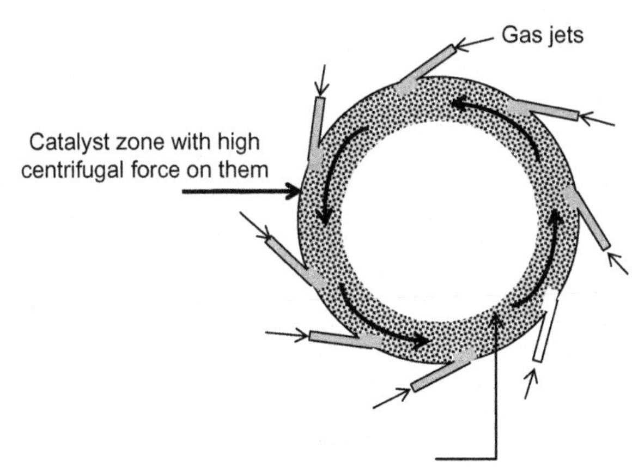

Figure 4.7 Static chamber rotating bed reactor.

have a finite life and become part of the product eventually. The catalysts and the nanomaterials are then removed from the reactor as a product. Purification, discussed next, can separate the two, but the raw form is referred in the rest of this book as "as-produced" GN or CNT.

4.3.4 Purity

FBRCVD can achieve very high-purity GN. We have discussed that the growth mechanism of GN involves growth from one or both faces of the catalyst particle. If the coking or other type of fouling dictates the premature termination of a production run, chances are the fibers have tangled around the catalyst particle. The removal of the catalyst from the final product in this case can be difficult, and may require harsh chemical treatment. If a production run is terminated while the catalysts still have relatively stable, but low activity/conversion, the catalyst particle will mostly be located at the end of the fiber. In this case if purification of the GN is required for any particular application, the location of the catalyst gives very easy access to the dilute acid solutions that can be used to dissolve the metal. High-energy dangling edges can cause stearic hindrance or simply cover the catalyst particle by forming tangled clusters. In most of the high-volume industrial uses envisaged, though, a small amount of catalyst in the product is not an issue, with the obvious caveat that the production run has yielded a high volume of GN. For example, for a yield of 250 $gGN/g_{catalyst}$, the concentration of the catalyst in the product is (0.4%), but if the yield is 50 $gGN/g_{catalyst}$, catalyst proportion increases to 2%, which may cross the tolerance threshold to prevent interferences from magnetic or electromagnetic forces, etc. If one evaluates the economics of synthesis, in any case, 50 $gGN/g_{catalyst}$ would not be viable for large-scale applications. So we can conclude that for the majority of the potential applications requiring large volumes of commercially viable GN today, the purity at 250 $g/g_{catalyst}$ is approximately the lower limit that needs to be exceeded.

4.3.5 Our experience with reactor design

My first exposure to nanomaterials was about 15 years ago, when I was in need of materials for an electrically activated catalysis process. It seemed at that time that the industry was on the cusp of explosive growth, and manufacturers will abound by the time we get to the large scale production point. We started experimenting with different materials to settle on GN. Much to our chagrin, no large-scale manufacturer emerged over the next two years. We licensed a catalyst and set out to design our own reactor.

4.3.5.1 Stage 1—Gravity FBCVD

Our first attempt to make these materials in large quantities was to utilize the gravity FBCVD concept. The reactor was a 5-in. diameter quartz tube within a tube furnace providing 3 temperature zones (see Fig. 4.8). The distributor was designed with orifices and a 10-μm ceramic filter disc. Each orifice was designed to generate a 0.5-psig pressure drop to generate proper distributed flow, though a 5 in. diameter does not

Figure 4.8 Gravity flow reactor—1 kg/day GN.

typically need such a precise distributor. The catalyst was Cu-Ni/MgO and came with operating instructions to run at 600–700°C, but no information on kinetics, growth rates, etc.

The reactor was run initially with 5 g of catalyst and gas flow that gave us an empty bed contact time of 2–3 minutes. From a starting phase of losing all the catalyst within minutes of start-up, it took us a year of modifications and adjustments with countless matrices of parameters to finally arrive at a design that gave us 12-hour runs per catalyst loading, and produced almost 100 $gGN/g_{catalyst}$. The runs were always terminated due to catalyst inactivation, by agglomerated GN that blocked access of the methane/carrier gas to the catalyst surface. This reactor design was later licensed to a Malaysian company where the diameter of the next reactor was to be 18 in. and flow characteristics, growth rates, yields in large void spaces were to be studied and eventually end up with a design of a gravity reactor with fractal distribution to create turbulence and enhance release of the fibers from the catalyst surface (see Fig. 4.9). You can probably guess that the project never materialized, and the larger reactor with fractal distribution never saw the light of day.

Back in the United States, we started studying the rotating FB reactors. The concept of a rotating chamber type FBCVD did not appeal to us, due to the mechanical complexity. We opted to try and design the static rotating reactor with tangential gas injection. We knew going in

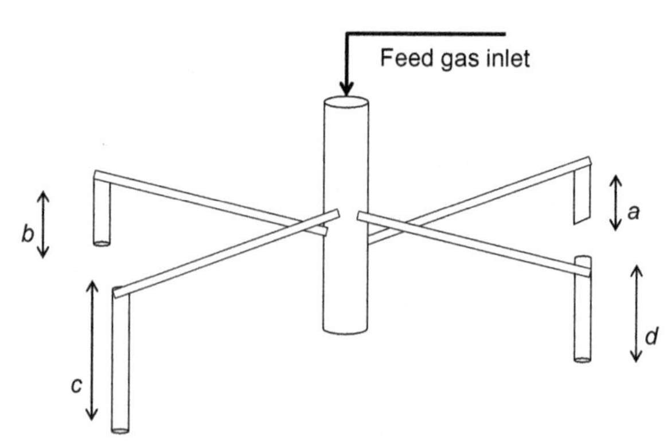

Figure 4.9 Conceptual fractal distributor: pressure drop across $a = b = c = d$, resulting in uniform flow rates. The gas is fed from the top of the feed pipe (also known as the "hub") and flows through the nozzles at uniform flow rates. Depending on the kind of flow regime required by the process, all lengths of *a,b,c,d* may be equal or different in size. In some processes multiple assemblies can be used to provide uniform physical as well as chemical environments.

that these reactors are mostly in pilot stages for CVD applications. We have tried many different catalysts successfully with this type of a pilot reactor. The graphitic degree and crystallinity of our GN material was the same or better than the previously licensed catalyst, leading us to believe a continuous removal of grown fibers allows better access to the nickel surface and consistent material can be synthesized. The nickel fraction in the catalyst seems to be the most important factor.

One of the main objectives with the reactor was to create the conditions where the GN would separate from the catalyst particles as well as prevent their agglomeration.

When I finally rode off into the sunset from my adventure in the nanomaterials field, the best results our team was able to obtain for long production runs ($\sim$300 gGN/$g_{catalyst}$) were from a static rotating FBCVD reactor. My advice to whoever wants to explore GN manufacturing is to look at rotating reactors. A conventional static rotating reactor is an excellent starting point for reactor design. For our team, the path to the final design from the standard design was short and easily achievable with some engineering and basic test parameter matrices, and an average IQ. A new, "patentable" reactor need not be designed, because static rotating FBCVD reactors work very well and are not complicated to build for small scale production of GN. Empirical data is always going to be necessary to design reactors. There are too many variables to consider in the theoretical engineering of an FBCVD reactor, regardles of whether the reactor type is single tube gravity, recirculating bed or rotating type. We found the published literature for the theoretical models and experimental data to be very helpful in shortening our time to establish precise operating parameters. The data we collected varied significantly from most of the publications but helped us understand the outer limits of the task we had at hand. We owe our gratitude to the many engineers and scientists who published such useful data.

Concluding the discussion on our efforts, we achieved our goal with a design that has tangential gas injection at three vertical points of the reactor, each with a different velocity based on empirical data for the growth of GN.

4.3.6 Health effects

The opposite is desired. There is a general concern that nanomaterials will be hazardous to humans. In our industrial and commercial world, we *fight* agglomeration by devising efficient methods of dispersion. Agglomeration and dispersion are important factors that have a direct bearing on the performance of materials we want to enhance with nanomaterials. We strive to reduce the level of agglomeration of nanomaterials in industrial uses such as enhancing the conductivity and mechanical properties of polymers. However, in a related universe of health care, the desire is the opposite. Agglomeration is beneficial for the minimization of the perceived health effects of nanomaterials. As commercialization progresses, there are bound to be major conflicts between the medical and industrial communities. Using nanomaterials in consumer products such as plastics will presumably have many limitations set by the health authorities. A consumer product can break, melt, or be crushed to expose nanomaterials that have simply been physically mixed with the base material, as by shear mixing. Nanomaterials that have been impregnated within the base material by chemical bonding would pose less danger, if any. More importantly, but easily solved, in large-scale manufacturing environments, the fear is that workers would be at an extreme risk from inhalation of these materials. A facemask to filter out nanomaterials would pretty much suffocate the person, unless he/she has the lungs of Superman/Superwoman. As a manufacturer, I don't believe manufacturing and packaging techniques are so archaic that manufacturers will not be able to deal with that scenario.

We all remember the asbestos crisis. The World Health Organization (WHO) fiber dimension limitations for respiration are: particles longer than 5 μm, <3 μm in diameter, and with an aspect ratio (length to diameter) $>3:1$ (WHO, 1981). This range certainly puts (as produced) nanomaterials out of the safety zone and presents a case for agglomeration being a favorable factor during production as well as for use in composites that would be subject to very high consumer exposure. Multiple inconclusive studies have been done to test for the carcinogenic effects of these materials [180–182], with and without agglomeration. The various results presented have so many generally unrelated test protocols or parameters, that a definitive interpretation was very difficult for a simple brain like mine. One parameter that was evident and mentioned was the inconsistency of fiber surface area of the samples they used. Sometimes these samples varied significantly from the manufacturer's given data. Another fact mentioned was the lower effects of acid-treated materials with $-COOH$ groups. The article did not clarify if this improvement resulted in acceptable levels or not. Surface areas of the fibers as well as their agglomerates were used as the most common variable for the dosage to be given to mice for pulmonary effects in these publications. It was found that SWCNTs agglomerate more than GN in general. There certainly is cause for caution, but more studies will be required to determine the parameters that will eventually become law. Industry would be wise to design and plan with that anticipated consequence. This fact adds to my argument of the timeline for nanomaterials to show up in mass produced products in the near future. In this case investors should be wise enough to invest with that eventuality in mind and therefore have realistic expectations.

This discussion further negates the use of graphene and graphene oxide with tiny particle sizes for industrial use. All nanomaterials will have to be used in a manner where the end use does not expose consumers to the original dimension material in the consumer products (a case for in situ polymerization, which I will discuss in chapter 6). We all know that even in the worst case scenario, consumer exposure will be extremely short term, and long-term exposure is what would probably be harmful, but try explaining that to the regulators.

CHAPTER FIVE

Costs of Manufacturing

5.1 CARBON NANOTUBES

Before we go into discussions about applications and their viability, let us arm ourselves with some costing knowledge about the three materials. We are pretty much familiar with the manufacturing and purification processes of graphitic nanofibers (GN), so I will just comment briefly on carbon nanotube (CNT) and graphene oxide (GO). Further down, we will look at the cost of converting graphite flakes to graphene. To keep our focus on practicality, especially for costs, we will skip the many, many references of CNT synthesis by carbon sources other than methane. These exotic gases have served their purpose well for the authors, but would be difficult if not impossible to scale their use as a feedstock for viability on the commercial end. For example, Moreno and Yoshimura [109] discuss hydrothermal production of multiwall carbon nanotubes (MWCNTs) from amorphous carbon at 800°C and 15,000 psig without any catalysts, Magrez et al. [184] talks about C_2H_2 as a feedstock with a low yield, and McCaldin et al. [183] discusses C_2H_4 as a feedstock. As we shall see, even using natural gas as a feedstock makes the products expensive to make. We have discussed the need and complexity of purification of CNTs. These procedures further increase production costs.

5.2 GRAPHENE OXIDE

Let me first explain how I have viewed GO in this book: The applications discussed in this book have been chosen with a purpose in mind that GN is an excellent candidate for the given applications. The research and publications cited with "graphene oxide" in their title describe flake graphite as the starting material and oxidation thereof. In my opinion, very few applications I studied would actually require few-layer graphene flakes, except when specifically noted. The materials

Graphitic Nanofibers.
DOI: http://dx.doi.org/10.1016/B978-0-323-51104-9.00005-4

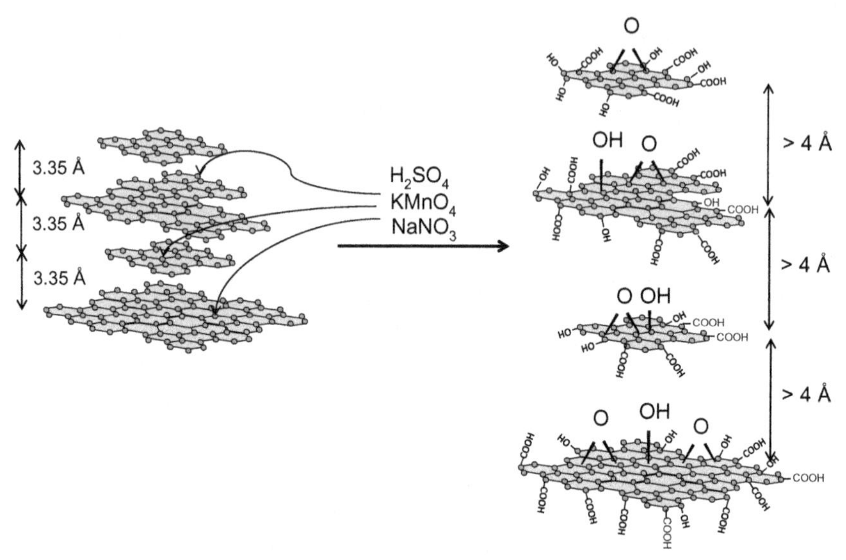

Figure 5.1 Oxidation of graphite particles with uneven lateral plane dimensions by the Hummers' method. The product has higher *d* spacing and is also called expanded graphite.

should really be called graphite oxide (GO). To maintain clarity, I will continue to use the acronym GO and use the word graphene oxide for the oxidized graphite. An overwhelming majority of references cite the procedure for GO synthesis by very harsh oxidation procedures, such as the Hummers and "slightly modified" Hummers (see Figs. 5.1 and 5.2).

The experimental details from the 1958 publication [113] outlining the Hummer's method are:

- 100 g of flake graphite
- 2.3 L 66 degrees Baume sulfuric (1800 g/L)
- 50 g sodium nitrate
- 300 g of potassium permanganate (1.9 gequiv.)
- 4.6 L of water followed by 14 L of water.

Though detailed explanations for the minor modifications to the Hummer's method are given in the experimental sections, papers claiming facile and cost effective large scale manufacturing pathways neglect to consider the environmental implications of toxic gas generation (NO_2, N_2O_4, ClO_2 (explosive)), waste treatment of the residual acids and oxidants, and the helacious cost of the chemicals required. Underscoring the purpose of the book, I emphasize the need to look at the cost of manufacturing these hyped, exotic materials. To start with, the original process generates

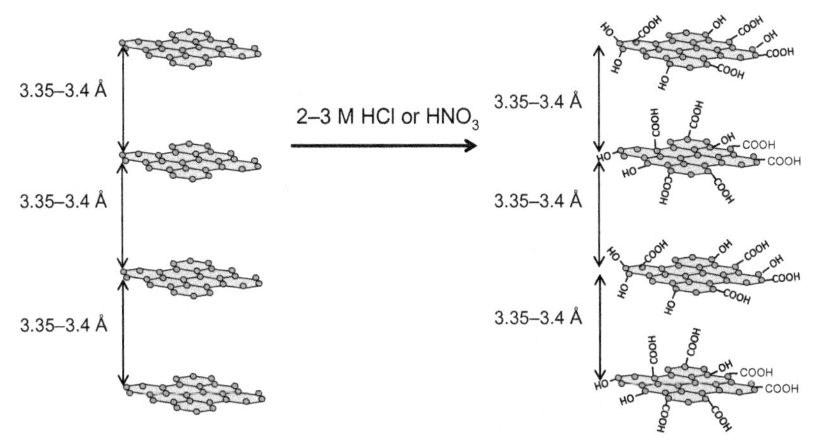

Figure 5.2 Platelet-type GNs are oxidized with mild acid. Unlike graphite, GN oxidize with no increase in *d* spacing. Only edges are functionalized, only —COOH & —OH groups found, do not get attacked on the basal plane.

oxides of nitrogen, which require treatment before exhaust. Graphite flakes can have many impurities that may vaporize with the NOx vapors. Since the nature of other gases in the exhaust will always be unknown, calculations for costing should be done for Selective Catalytic Reduction (SCR). The reducing agent is usually NH_3 and is injected at a 1:1 molar ratio:

$$2NO + 2NH_3 = 2N_2 + 3H_2O$$

The sodium nitrate used in the reaction (as a catalyst) will be the source of this NO. So let's see how much NH_3 would be required for the referenced batch of 100 grams of flake graphite. Hummer's recipe calls for 50 grams or ~0.6 moles of $NaNO_3$. We would therefore need 0.6 moles of NH_3 or ~10 grams of NH_3 on 100% basis. This cost is incorporated in the calculations to follow.

5.3 COST CALCULATIONS

The itemized costs of production for six products are tabulated in Table 5.1. As a reference point, the pricing of materials were obtained on September 30th, 2016. For commodities, I have taken the world spot prices and added a conservative 50% to arrive at the final price for delivered product, as given in Table 5.1. If a company is large enough to buy at world trading commodity prices, they will also have a much

Table 5.1 Summary of costs and price per $kg_{product}$

	GN-Biogas	GN-Syngas	GN-NG	CNT-Biogas	CNT Syngas	CNT-NG	GO
Direct operating cost							
Cost of feed gas per kg	–	–	$3.39	–	–	$8.40	–
Cost of gas purification per kg	$2.96	$2.96	$0.62	$7.60	$7.60	$1.60	–
Catalyst cost	$3.33	$3.33	$3.33	$9.50	$9.50	$9.50	–
HNO_3 cost	–	–	–	$22.31	$22.31	$22.31	–
HCl cost	$1.32	$1.32	$1.32	–	–	–	–
NaOH cost	$1.20	$1.20	$1.20	$4.95	$4.95	$4.95	$19.95
H_2SO_4 cost	–	–	–	–	–	–	$14.22
$NaNO_3$ cost	–	–	–	–	–	–	$0.21
$KMnO_4$ cost	–	–	–	–	–	–	$6.30
DI water cost	$0.24	$0.24	$0.24	$0.76	$0.76	$0.76	$0.56
Labor cost	$1.68	$1.68	$1.68	$1.68	$1.68	$1.68	$1.68
Payroll expense	$0.20	$0.20	$0.20	$0.20	$0.20	$0.20	$0.20
Maintenance	$0.34	$0.40	$0.29	$0.86	$0.91	$0.69	$0.23
NO_x abatement	–	–	–	–	–	–	$0.04
Flake graphite	–	–	–	–	–	–	$10.50
Packaging (estimated)	$1	$1	$1	$1	$1	$1	$1
Total direct operating costs	**$12.27**	**$12.33**	**$13.27**	**$48.86**	**$48.91**	**$51.09**	**$54.89**

Other operating expenses							
Insurance	$0.34	$0.40	$0.29	$0.86	$0.91	$0.69	$0.23
Management salaries	$1.43	$1.43	$1.43	$1.43	$1.43	$1.43	$1.43
Payroll taxes + benefits	$0.17	$0.17	$0.17	$0.17	$0.17	$0.17	$0.17
QC reject	$0.61	$0.62	$0.66	$2.44	$2.45	$2.55	$2.74
Loan payments	$2.18	$2.55	$1.82	$5.46	$5.82	$4.36	$1.45
Depreciation	$1.71	$0.00	$1.43	$4.29	$4.57	$3.43	$1.14
Misc. overheads plant related	$1.23	$1.23	$1.33	$4.89	$4.89	$5.11	$5.49
Total indirect operating costs	$7.67	$6.40	$7.13	$19.54	$20.24	$17.74	$12.65
Total cost to Mfr. per kg$_{product}$	**$19.94**	**$18.73**	**$20.40**	**$68.40**	**$69.15**	**$68.83**	**$67.54**
Estimating OEM (original equipment manufacturer) price							
Mfr. typical net earnings	15.00%	15.00%	15.00%	15.00%	15.00%	15.00%	15.00%
Effective corporate tax rate	30.00%	30.00%	30.00%	30.00%	30.00%	30.00%	30.00%
Earnings before taxes (EBIT) (as % of revenue)	50.00%	50.00%	50.00%	50.00%	50.00%	50.00%	50.00%
Total cost before depreciation	$18.23	$18.73	$18.97	$64.11	$64.58	$65.40	$66.40
Earnings before depreciation (EBITDA)	$18.23	$37.46	$37.91	$128.22	$129.16	$130.80	$132.80
Revenues (Cost + EBITDA)	$36.46	$56.19	$56.91	$192.33	$193.74	$196.20	$199.20
Mfr. sale price to distributors	$36.46	$41.46	$40.80	$136.80	$138.30	$137.66	$135.00
Distributor sale price to customer	$47.40	$53.90	$53.04	$177.84	$179.90	$178.96	$175.60
Avg. price-per kg$_{product}$		**$51**			**$179**		**$176**

larger overhead factor in the costing. These two should cancel each other out.

The input values and detailed calculations of the tabulated costs in Table 5.1 are presented at the end of this chapter.

Table 5.1 lists the values of individual cost components and final cost per Kg of each material for several feed source options.

- Manufacturing GN with LFG as the carbon source.
- Manufacturing GN with AD biogas as a carbon source.
- Manufacturing GN with natural gas, including purification required and losses to achieve a Methane Number of 100%.
- Manufacturing CNT with LFG as the carbon source.
- Manufacturing CNT with AD biogas as a carbon source.
- Manufacturing CNT with natural gas, including purification required and losses to achieve a Methane Number of 100%.
- Manufacturing GO.

5.4 MAKING GRAPHENE

Let's look at what it would cost to make graphene, now that we know what GO costs per Kg. Synthesizing large volumes of graphene requires GO or some other form of graphitic nanomaterial. Graphene was first isolated and studied by a mechanical "peeling" of graphite using an adhesive tape. Graphene layers up to a few millimeter in size can be produced by peeling. This technique is currently (and likely to remain) the source for the small demand of basic research. Publications describe new methods of large-area graphite synthesis, including CVD and direct growth on catalysts [185–192]. This method is not practical for large-volume production. The most widely referenced method of synthesizing graphene flakes is via reduction of GO made from flake graphite. A case can be made for GN to be the starting material for synthesizing graphene. Ribbon-type GN would be the best candidates for exfoliation to graphene. The basal planes would stay pristine, and 1 to 10-μm graphene flakes could be synthesized. Exfoliation techniques are worth a look from a cost perspective for ribbon-type GN to graphene. A cursory look at the costs associated with exfoliation to obtain graphene is given below.

5.4.1 Sonication

Ultrasonication is used extensively in research labs. As the name implies, acoustic energy replaces thermal, electrical, and other energy forms. The process is also used for emulsification of immiscible liquids to provide energy for homogeneous chemical reactions and some other niche applications. It can be used in providing the energy required to overcome the van der Waals forces and cleave expanded graphitic plates into single-layer single crystal type graphene particles. Ultrasound has been used to exfoliate expanded graphite (EG) flakes in water, solvents, and surfactants [193–195]. If we consider using functionalized GN in aqueous solution or as-produced in surfactant solutions, rather than using graphite flakes, we could gain some benefits that GN offer over graphite flakes: (1) Consistent characteristics for each graphene plane. (2) Higher purity. (3) Higher crystallinity. Using GN in place of graphite flakes could provide better-quality graphene particles with better yields. The length of the ribbon-type GN would dictate the size of the graphene flakes. These advantages would come at the cost of a slightly higher energy consumption since GN would not be "expanded" as most GO-treated graphite flakes are, and the entire VdW force would have to be overcome.

For cost calculations, unfortunately, Refs. [193–195] do not extrapolate the power consumption of this technique per unit mass, nor do they specify cost of other materials. Six hundred watts of ultrasonic power was used for 5 minutes to exfoliate 1 g of EG to "20-layer" graphene flakes, which the authors term FLG (few-layer graphene) [193]. The two-step process also requires heating the EG to 1100°C prior to sonication. It is not clear how the authors acquired EG, or what process was used. On a manufacturer's website (Hielscher), some estimates of volume processed per day per KW of sonic power are given for graphene concentrations of 1–0.001%. In fact, most other references I have read on the use of ultrasonics for exfoliation of graphene plates also use very dilute solutions of either GO or graphite suspensions to evaluate the ability of ultrasound to exfoliate. In Table 5.2, I have tried to extrapolate the values buried in the texts. The table also lists if the experiments were done on graphite flakes GO (sometimes referred to as EG). These costs may seem high due to the small amounts of graphene recovered in the experiments. But the concentrations used must be modified to adjust these costs.

Table 5.2 Derivation of direct costs for exfoliation by sonication

References	Grams graphene recovered	Power (KW-h)	Solvent (mL)	Total power ($)	Total solvent ($)	Cost per $kg_{product}$ ($)
[198]	2.0 from graphite	9.00	700	0.90	1.40	1150
[193]	2.3 from graphite	17.63	700	1.77	1.40	1378
[201]	1.3 from graphite	1	1000	0.10	1.0	846

Costing Assumptions:
1. Electrical power to ultrasound energy conversion has been assumed to be 60% [197].
2. Cost of electricity is assumed to be US$0.10 per KW-h.
3. Cost of solvents is assumed to be US$2 per kg with a specific gravity ~1.0.

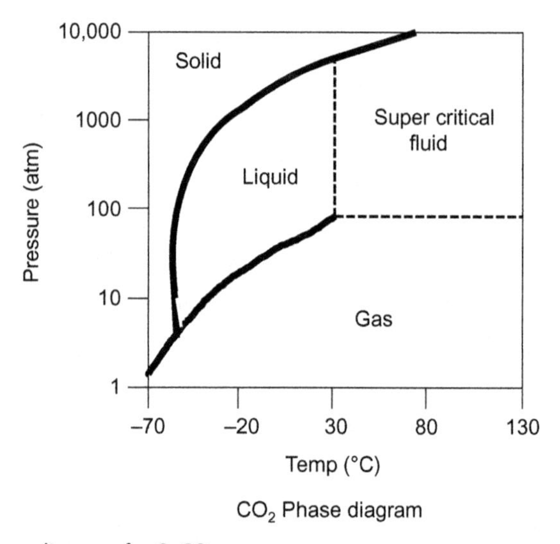

Figure 5.3 Phase diagram for $ScCO_2$.

5.4.2 Supercritical carbon dioxide

Another method for a non-GO route exfoliation is by using supercritical carbon dioxide [196]. In its critical phase, CO_2 behaves neither like a liquid nor a gas (see Fig. 5.3). $ScCO_2$ has a surface tension that is an order of magnitude lower than propane, and a very low viscosity. It penetrates tiny pores or cracks and mixes readily with oils. These qualities result in unique solvent-like qualities. $ScCO_2$ has been used widely for extraction of oils, enhanced oil recovery, etc. Once the material to be extracted is dissolved in the $ScCO_2$, the pressure and/or temperature conditions are changed, at which point the CO_2 releases as a gas, and the solvate can be recovered. Again, by avoiding the GO route, intercalation and exfoliation by $ScCO_2$ has the possibility of producing pristine graphene planes for graphite, or consistent sized platelets from ribbon-type GN. Li et al. [196] describe complete depressurization of the $ScCO_2$ for the rapid exfoliation of the basal planes.

The process equipment for such a route on an industrial scale would be expensive though it has been used by the oil industry for decades now. High-pressure vessels, multistage compressors with heat removal between stages, loss of $ScCO_2$ due to maintenance of gas seals, special material requirements for O-rings to resist explosive decompression (CO_2 gets trapped in the materials), and last but not least, high-purity CO_2 as a feedstock are some of the reasons. Power consumption on a large scale may be calculated by using atmospheric intake, and a 1800-psig discharge a three-stage compressor, and a power consumption cost at (US$0.10 per KW-h). This gives us an estimated electrical cost of *~$80 per kg*. An analysis similar to the one we did on the GN and CNTs, with the proper equipment costs would give more accurate numbers. Further investigation with pilots should be carried out to find the actual numbers.

As is common knowledge, none of the above methods has been tested or implemented on a large-scale commercial exfoliation of graphite/ graphene oxide.

5.5 CALCULATIONS FOR TABLE 5.1

Table 5.3 Commodity chemical prices

Chemical	**World commodity price**	**Small manufacturer's price 50% higher**
	Cost/kg US$	**Cost/kg US$**
Flake graphite	10	15
H_2SO_4	0.32	0.48
$KMnO_4$ (100%)	2	3
$Na_2O_5S_2$ (100%)	1.6	2.4
NaOH (100%)	0.5	0.75
$NaNO_3$ (100%)	0.4	0.6
NH_3 (anhydrous)	0.18	0.27
HCl 31%— muriatic	0.21	0.315
HNO_3 (16M)	1.1	1.65
Catalyst for GN	600	900
Hydrazine hydrate	3	4.5

Table 5.4 Operating cost inputs

Other operating variables	**Value**	**Unit**
Unit cost of natural gas	15	\$/MMBTU
Unit cost of natural gas treatment	0.1	\$/nM^3
Unit cost biogas treatment	0.25	\$/nM^3
Conversion rates	35	%
Selectivity to CNT	35	%
Selectivity to GN	90	%
CH_4 content in biogas	50	%
Bulk density flake graphite	0.667	g/cc
Bulk density CNT and GN	0.300	g/cc
Unpurified product	300	g/g$_{catalyst}$
Plant capacity	50	kg/h
Plant capacity	1000	kg/day
Operating hours	20	Per day
Days/year	350	Days
Capacity	350,000	kg/year
Unit labor rate	35	\$/h
Maintenance	2	% of capex/year
Cumulative mgmt. salaries	500, 000	US\$/year
Payroll expenses	12	% of gross pay
Insurance	3	% of capex year
QC rejects	5	% (To give net plant cap).
Concentration of HCl	3	Normal
Concentration of NaOH used	100	% by weight
Concentration of HNO_3 used	16	Molar
Concentration of H_2SO_4 used	98	% by weight
Baseline quantity of total carbon deposited	300	g/g$_{catalyst}$

Table 5.5 Financial cost inputs

Financial factors	
Capex GN biogas	\$6,000,000
Capex GN syngas	\$7,000,000
Capex GN NG	\$5,000,000
Capex CNT biogas	\$15,000,000
Capex CNT syngas	\$16,000,000
Capex CNT NG	\$12,000,000
Capex GO	\$4,000,000
Loan term	84 months
Interest rate	8%
Leverage	70%
Depreciation	10-year straight line

Table 5.6 Prepurification quantity of product required to achieve desired output

Weight of raw product required before purification to achieve rated capacity	
Weight of produced material required to achieve 1000 $kg_{GN-product}$	$= \frac{1000\ g_{total\ GN}\ \text{required}}{90\%\ \text{selectivity}}$
	$= 1.11\ kg_{raw\ product}/kg_{GN}$
Weight of produced material requiring purification for 1 $kg_{CNT-product}$	$= \frac{1000\ g_{total\ product}\ \text{desired}}{35\%\ \text{selectivity}}$
	$= 3\ kg_{raw\ product}/kg_{CNT}$
Weight of flake graphite required to make 1 $kg_{GO-product}$	$= \frac{1\ kg_{GO\ product}}{180\%\ \text{due to } 80\%\ \text{weight gain}}$
	$= 0.56\ kg_{flake}/kg_{GO}$
Weight of flake graphite required to make 1 $kg_{GO-product}$ including waste	$= 556\ g_{flake\ graphite} \times 125\%$ for attrition and multiple rinsing losses
	$= 0.70\ kg_{flake}/kg_{GO}$

Table 5.7 Biogas/syngas requirements for GN production

Number of kg-moles of product per day	$= \frac{1000\ \text{kg}}{12\ \text{kg/kg} - \text{mol}}$
	$= 83\ \text{kg} - \text{mol/day}$
Yield	$= 35\%$ conversion $\times$ 90% selectivity
	$= 31.50\%$
Methane required (mol/day)	$= \frac{83\ \text{kg} - \text{mol}}{31.50\%\ \text{yield}}$
	$= 265\ \text{kg} - \text{mol/day}$
Methane required (nM^3/day)	$= \frac{265\ \text{kg} - \text{mol methane} \times 22.40\ nM^3/\text{mol}}{\text{at STP (assuming Ideal gas law)}}$
	$= 5926\ nM^3/\text{day}$
Biogas required	$= \frac{5926\ nM^3/\text{day pure methane}}{50\%\ \text{methane content by volume}}$
	$= 11,852\ nM^3/\text{day}$
Biogas required per $kg_{product}$ (nM^3)	$= \frac{11,852\ nM^3\text{biogas/day}}{1000\ \text{kg/day}}$
	$= 11.9\ nM^3/\text{day}$
Biogas per hour	$= \frac{11,852\ nM^3\text{biogas/day}}{24\ \text{h/day}}$
	$= 494\ nM^3/\text{h}$

Table 5.8 Natural gas requirement for GN production

Methane volume/day from (biogas calculation)	$= \frac{5926\ nM^3/\text{day methane}}{95\%\text{purity by volume}}$
	$= 6238\ nM^3/\text{day}$
NG volume per kilogram	$= \frac{6238\ nM^3/\text{day}}{1000\ \text{kg/day}}$
	$= 6.2\ nM^3/kg_{GN}$
NG volume per hour	$= \frac{6238\ nM^3/\text{day}}{24\ h/\text{day}}$
	$= 260\ nM^3/h$
Cubic feet NG required	$= 6.2\ nM^3/\text{kg} \times 35.3\ \text{CF}/nM^3$
	$= 219\ \text{CF}/kg_{GN}$
BTUs value of gas	$= 219\ \text{CF} \times 1000\ \text{BTUs/CF}$
	$= 219,000\ \text{BTUs}$
Cost of gas per $kg_{product}$	$= 219,000\ \text{BTUs/kg} \times \$15$ per million BTUs
	$= \$3.39$ per kg_{GN}

Table 5.9 Biogas/syngas required for CNT production

Number of kg – moles of CNT required per day	$= \frac{1000\text{ kg}}{12\text{ kg/kg} - \text{mol}}$
	$= 83\text{ kg} - \text{mol/day}$
Yield	$= 35\%\text{ conversion} \times 35\%\text{ selectivity}$
	$= 12.25\%$
Methane required (mol/day)	$= \frac{83\text{ kg} - \text{mol}}{12.25\%\text{ yield}}$
	$= 678\text{ kg} - \text{mol/day}$
Methane required (nM^3/day)	$= 678\text{ kg} - \text{mol methane} \times 22.40\text{ nM}^3/\text{mol}$ at STP (assuming Ideal gas law)
	$= 15,187\text{ nM}^3/\text{day}$
Biogas required	$= \frac{15,187\text{ nM}^3/\text{day pure methane}}{50\%\text{ methane content by volume}}$
	$= 30,374\text{ nM}^3/\text{day}$
Biogas required per $\text{kg}_{\text{product}}$ (nM^3)	$= \frac{30,374\text{ nM}^3\text{ biogas/day}}{1000\text{ kg/day}}$
	$= 30.4\text{ nM}^3/\text{kg}_{\text{GN}}$
Biogas per hour	$= \frac{30,374\text{ nM}^3\text{ biogas/day}}{24\ h/\text{day}}$
	$= 1266\text{ nM}^3/h$

Table 5.10 Natural gas required for CNT manufacturing

Methane volume per day	$= \frac{15,187\text{ nM}^3/\text{day methane}}{95\%\text{ purity by volume}}$
	$= 15,986\text{ nM}^3/\text{day}$
NG volume per kilogram	$= \frac{15,986\text{ nM}^3/\text{day}}{1000\text{ kg/day}}$
	$= 16.0\text{ nM}^3/\text{kg} - \text{GN}$
NG volume per hour	$= \frac{15,986\text{ nM}^3/\text{day}}{24\ h/\text{day}}$
	$= 666\text{ nM}^3/h$
Cubic feet NG required	$= 16.0\text{ nM}^3/\text{kg} \times 35.3\text{ CF/nM}^3$
	$= 565\text{ CF/kg}_{\text{GN}}$
BTUs value of gas	$= 565.0\text{ CF} \times 1000\text{ BTUs/CF}$
	$= 565,000\text{ BTUs}$
Cost of gas per $\text{kg}_{\text{product}}$	$= 565,000\text{ BTUs/kg} \times \$15\text{ per million BTUs}$
	$= \$8.48\text{ per kg}_{\text{GN}}$

Table 5.11 Cost of catalyst for GN production

Total carbon deposited	$= 300\ g/g_{catalyst}$
Yield of GN ($g/g_{catalyst}$)	$= 300\ g_{carbon}/g_{catalyst} \times 90\%$ selectivity
	$= 270\ g_{GN}/g_{catalyst}$
Catalyst weight per kg_{GN}	$= 3.33\ g_{catalyst}$ consumed
Catalyst cost	$= 3.33\ g_{catalyst}$ consumed $\times$ \$1000/1000 g
	$=$ \$3.33/kg_{GN}

Table 5.12 Cost of catalyst for CNT production

Total carbon deposited	$= 300\ g/g_{catalyst}$
Yield of CNT ($g_{CNT}/g_{catalyst}$)	$= 300\ g/g_{catalyst} \times 35\%$ selectivity
	$= 105\ g/g_{catalyst}$
Catalyst weight consumed per kilogram	$= 9.5\ g_{catalyst}$ consumed
Catalyst cost	$= 9.5\ g_{catalyst}$ consumed $\times$ \$1000/1000 g
	$=$ \$9.50/kg_{CNT}

Table 5.13 Volume of acids for purification

Volume of acids required to soak raw materials for purification—1 bed volume	
Bulk density of CNT-GN	300 g/L
Weight per 1 m^3—CNT-GN	300 kg/m^3
Weight of produced material required to yield 1000 $kg_{CNT\text{-}product}$	2900 kg/day-CNT
Volume of HNO_3 required for purification to achieve 1000 $kg_{CNT\text{-}product}$	9.67 m^3
Weight of produced material required to achieve 1000 $kg_{GN\text{-}product}$	1111 kg
Volume of HCl required to purify GN for 1000 $kg_{product}$	3.7 m^3

Table 5.14 Cost of purification of GN

Purification–oxidation of GN	
Weight of (31%) HCl used (kg)	$= 3703\ L_{HCl}/\text{day} \times 1.15$ specific gravity
	$= 4$ kg/day
Weight of HCl ($kg/kg_{product}$)	$= \frac{4259\ kg_{HCl}/\text{day}}{1000\ kg/\text{day}}$
	$= 4.26\ kg/kg_{GN}$
Cost of HCl (31%) ($/kg)	$= 4.26\ kg_{HCl}/kg_{product} \times \$0.31$ per $kg_{commodity\ price}$
	$= \$1.32$ per kg_{GN}

Table 5.15 Cost of purification of CNTs

Purification of CNTs	
Weight of (15.8 M) HNO_3 used (kg_{HNO_3})	$= 9524\ L_{HNO_3}/\ \text{day} \times 1.42$ specific gravity
	$= 13,524$ kg/day
Weight of HNO_3 ($kg/\ kg_{product}$)	$= \frac{13,524\ kg_{HNO_3}/\text{day}}{1000\ kg/\text{day}}$
	$= 13.500\ kg/kg_{CNT}$
Cost of HNO_3 (15.8 M) ($/kg)	$= 13.500\ kg_{HNO_3}/kg_{product} \times \$1.65$ per $kg_{commodity\ price}$
	$= \$22.31$ per kg_{CNT}

Table 5.16 Cost of chemicals used for GO synthesis

Cost of H_2SO_4	$= 29.62$ kg $_{H_2SO_4}/kg_{product}$ $H_2SO_4 \times \$0.48$ per kg 98% H_2SO_4 $\times \$0.48$ per kg 98% H_2SO_4
	$= \$14.22$
Weight of $NaNO_3$(g/kg_{GO})	$= 700.000\ g_{flake\ graphite} \times 0.5(50\ g_{NaNO_3}/100\ g_{flake\ graphite})$
	$= 350$
Cost of $NaNO_3$ per kg_{GO}	$= 350\ g_{NaNO_3}$
	$= \$0.21$
Weight of $KMnO_4$ per(g/kg_{GO})	$= 700\ g_{flake\ graphite} \times 2100\ g_{KMnO_4}$
	$= 2100$
Cost of $KMnO_4$	$= 2100\ g_{KMnO_4}$
	$= \$6.30$

Table 5.17 Cost of wastewater treatment for GN purification

Cost of NaOH to neutralize HCl used for GN purification	
Moles of HCl to neutralize (maximum) per day	$= 6238\ L_{HCl}/day \times 9.8\ mol/L_{HCl}$
	$= 61,132\ mol/day$
Moles of HCl to neutralize (maximum) per kilogram	$= \dfrac{61,132\ mol/day}{1000\ kg_{product}/day}$
	$= 61.1\ mol/kg_{GN}$
Moles of NaOH required per $kg_{product}$	$= 61\ mol\ HCl/kg_{product} \times 1.1\ mol\ NaOH/mol\ HCl$—10% excess
	$= 67\ mol$
Weight of NaOH required as 100% per $kg_{product}$	$= 67\ mol\ NaOH/kg \times 40\ g/mol$ NaOH molecular weight
	$= 2690\ g/kg_{GN}$
Cost of NaOH to neutralize HCl per $kg_{product}$	$= 2690\ g_{NaOH} \times \$0.75$ per kg 100% NaOH
	$= \$2.02$ per kg_{GN}

Table 5.18 Cost of wastewater treatment for CNT purification

Moles of HNO_3 to neutralize (maximum) per day	$= 9524\ L_{HNO_3}/\text{day} \times 15.8\ \text{mol}/L_{HNO_3}$
	$= 150,479\ \text{mol/day}$
Moles of HNO_3 to neutralize (maximum) per kilogram	$= \dfrac{150,479\ \text{mol/day}}{1000\ \text{kg}_{\text{product}}/\text{day}}$
	$= 150\ \text{mol/kg}_{\text{CNT}}$
Moles of NaOH required per $\text{kg}_{\text{product}}$	$= 150\text{mol}\ HNO_3/\text{kg}_{\text{product}} \times 1.1\ \text{mol NaOH/mol}\ HNO_3$—10% excess
	$= 165\ \text{mol}$
Weight of NaOH required as 100% per $\text{kg}_{\text{product}}$	$= 165\ \text{mol NaOH/kg}_{\text{product}} \times 40\ g/\text{mol}$ NaOH molecular weight
	$= 6600\ g/\text{kg}_{\text{CNT}}$
Cost of NaOH to neutralize HCl per $\text{kg}_{\text{product}}$	$= 6600\ \text{g}_{\text{NaOH}} \times \$0.75$ per kg 100% NaOH
	$= \$4.95$ per kg_{CNT}

Table 5.19 Cost of wastewater treatment from GO synthesis

Cost of NaOH required to neutralize H_2SO_4 used in GO synthesis	
Kilograms of H_2SO_4 to neutralize (maximum) per day	$= 16,100\ L_{H_2SO_4}/day \times 1.84$ specific gravity of 98% H_2SO_4
	$= 29,624$
Kilograms of H_2SO_4 to neutralize (maximum) per $kg_{product}$	$= \frac{29,624\ kg/day}{1000\ kg_{product}/day}$
	$= 29.62$
Moles of H_2SO_4 to neutralize per $kg_{product}$	$= \frac{29.62 kg_{H_2SO_4}/kg_{GO}}{98\ g/mol\ H_2SO_4}$
	$= 302$
Moles of NaOH required per $kg_{product}$	$= 302$ mol $H_2SO_4/kg_{GO} \times 2.2$ mol NaOH per mole H_2SO_4—10% excess
	$= 664$
Weight of NaOH required as 100% per $kg_{product}$	$= 665$ mol $NaOH/kg_{GO} \times 40$NaOH molecular weight
	$= 26,600$
Cost of NaOH to neutralize H_2SO_4 per $kg_{product}$	$= 26,600\ g_{NaOH} \times \$0.75$ per kg 100% NaOH
	$= \$19.95$

Table 5.20 Cost of deionized water for rinsing products after chemical treatment

DI water required for GO (L)	$= 700\ g_{\text{flake graphite}}/kg_{GO} \times 0.2\ (20\ L/100\ g_{\text{flake graphite}})$
	$= 140$ L
DI water cost per kg_{GO}	$= 140\ L_{\text{DI water}}/kg_{GO} \times 0.004$ (US\$ $4/m^3$—operating cost of RO/DI system)
	= \$0.56 per kg_{GO}
DI water required for CNT	$= 10\ L_{\text{raw product}}/kg_{CNT} \times 20$ bed volumes DI water to rinse acid + fines
	$= 190\ L/kg$ CNT
DI water cost per kg_{CNT}	$= 190\ L/kg_{CNT} \times 0.004$ (US\$ $4/m^3$—operating cost of RO/DI system)
	= \$0.76per kg_{GN-CNT}
DI water required for GN	$= 3\ L_{\text{raw product}}/kg_{GN} \times 20$ bed volumes of DI water
	$= 60\ L/kg_{GN}$
DI water cost per kg_{CNT}	$= 60L_{\text{raw product}}/kg_{GN} \times 0.004$ (US\$ $4/m^3$—operating cost of RO/DI system)
	= \$0.24 per kg_{GN}

Table 5.21 Cost of NO_x abatement from GO synthesis
NOx abatement for GO $2NO + 2NH_3 = 2N_2 + 3H_2O$

Anhydrous NH_3 required per kg_{GO}	$= 700\ g_{flake\ graphite} \times 0.2\ (10\ g_{NH3}/50 g_{NaNO_3})$
	$= 140\ g/kg_{GO}$
Cost of ammonia	$= 140\ g/kg_{GO} \times 0.00053$ (US\$ 0.27/kg)
	$= \$0.04$ per kg_{GO}

Table 5.22 Labor costs

Number of labor hours	$= 2$ people $\times$ 24 h
	$= 48$
Labor cost per day	$= 48\ h/\text{day} \times \$35$ (hourly wage)
	$= \$1680$
Labor cost per $kg_{product}$	$= \dfrac{\$1680 \text{ per day}}{1000\ \text{kg/day}}$
	$= \$\ 1.68$

Table 5.23 Maintenance costs for GN plant configurations

Annual cost of GN biogas plant	$= \$6,000,000$ Capex $\times$ 2% rule of thumb estimate
	$= \$120,000$
Cost per kg_{GN} from biogas	$= \dfrac{\$120,000 \text{ annual cost}}{350,000\ kg_{GN} \text{ produced}}$
	$= \$0.34$
Annual cost of GN – syngas	$= \$7,000,000$ Capex $\times$ 2% rule of thumb estimate
	$= \$140,000$
Cost per kg_{GN} from syngas	$= \dfrac{\$140,000 \text{ annual cost}}{350,000\ kg_{GN} \text{ produced}}$
	$= \$0.40$
Annual cost of GN NG plant	$= \$5,000,000$ Capex $\times$ 2% rule of thumb estimate
	$= \$100,000$
Cost per kg_{GN} from NG	$= \dfrac{\$100,000 \text{ annual cost}}{350,000\ kg_{GN} \text{ produced}}$
	$= \$0.29$

Table 5.24 Maintenance costs for CNT plant configurations

Maintenance costs for respective CNT plants	
Annual cost of CNT biogas plant	$= \$15,000,000$ Capex $\times$ 2% rule of thumb estimate
	$= \$300,000$
Cost per kg_{CNT} from biogas	$= \dfrac{\$300,000 \text{ annual cost}}{350,000\ kg_{GN} \text{ produced}}$
	$= \$0.86$
Annual cost of CNT – syngas	$= \$16,000,000$ Capex $\times$ 2% rule of thumb estimate
	$= \$320,000$
Cost per kg_{CNT} from syngas	$= \dfrac{\$320,000 \text{ annual cost}}{350,000\ kg_{GN} \text{ produced}}$
	$= \$0.91$
Annual cost of CNT NG plant	$= \$12,000,000$ Capex $\times$ 2% rule of thumb estimate
	$= \$240,000$
Cost per kg_{CNT} from NG	$= \dfrac{\$240,000 \text{ annual cost}}{350,000\ kg_{GN} \text{ produced}}$
	$= \$0.69$

Table 5.25 Maintenance cost for GO plant

Annual cost of GO	$= \$4,000,000$ Capex $\times$ 2% rule of thumb
	$= \$80,000$
Cost per kg_{GO}	$= \dfrac{\$80,000 \text{ annual cost}}{350,000\ kg_{GN} \text{ produced}}$
	$= \$0.23$

Table 5.26 Insurance costs

Annual cost of GN biogas plant	$= \$6,000,000$ Capex $\times$ 3% rule of thumb
	$= \$120,000$
Cost per kg_{GN} from biogas	$= \dfrac{\$120,000 \text{ annual cost}}{350,000\ kg_{GN} \text{ produced}}$
	$= \$0.34$ per $kg_{GN-biogas}$
Annual cost of GN – syngas	$= \$7,000,000$ Capex $\times$ 3% rule of thumb
	$= \$140,000$
Cost per kg_{GN} from syngas	$= \dfrac{\$140,000 \text{ annual cost}}{350,000\ kg_{GN} \text{ produced}}$
	$= \$0.40$ per $kg_{GN-syngas}$
Annual cost of GN – NG plant	$= \$5,000,000$ Capex $\times$ 3% rule of thumb
	$= \$100,000$
Cost per kg_{GN} from NG	$= \dfrac{\$100,000 \text{ annual cost}}{350,000\ kg_{GN} \text{ produced}}$
	$= \$0.29$ per kg_{GN-NG}
Annual cost of CNT biogas plant	$= \$15,000,000$ Capex $\times$ 3% rule of thumb
	$= \$300,000$
Cost per kg_{CNT} from biogas	$= \dfrac{\$300,000 \text{ annual cost}}{350,000\ kg_{GN} \text{ produced}}$
	$= \$0.86$ per $kg_{CNT-biogas}$
Annual cost of CNT – syngas	$= \$16,000,000$ Capex $\times$ 3% rule of thumb
	$= \$320,000$
Cost per kg_{CNT} from syngas	$= \dfrac{\$320,000 \text{ annual cost}}{350,000\ kg_{GN} \text{ produced}}$
	$= \$0.91$ per $kg_{CNT-syngas}$
Annual cost of CNT NG plant	$= \$12,000,000$ Capex $\times$ 3% rule of thumb
	$= \$240,000$
Cost per kg_{CNT} from NG	$= \dfrac{\$240,000 \text{ annual cost}}{350,000\ kg_{GN} \text{ produced}}$
	$= \$0.69$ per kg_{CNT-NG}
Annual cost of GO	$= \$4,000,000$ Capex $\times$ 3% rule of thumb
	$= \$80,000$
Cost per kg_{GO}	$= \dfrac{\$80,000 \text{ annual cost}}{350,000\ kg_{GN} \text{ produced}}$
	$= \$0.23$ per kg_{GO}

Table 5.27 Depreciation expenses

Annual depreciation of GN biogas plant	$= \frac{\$6,000,000 \text{ Capex}}{10 \text{ years}}$
	$= \$600,000$
Depreciation per kg_{GN} from biogas	$= \frac{\$600,000 \text{ annual cost}}{350,000\ kg_{GN} \text{ produced}}$
	$= \$1.71$ per $kg_{GN-biogas}$
Annual depreciation of GN – syngas	$= \$7,000,000$ Capex $\times$ 10 years
	$= \$700,000$
Depreciation per kg_{GN} from syngas	$= \frac{\$700,000 \text{ annual cost}}{350,000\ kg_{GN} \text{ produced}}$
	$= \$2.00$ per $kg_{GN-syngas}$
Annual depreciation of GN – NG plant	$= \$5,000,000$ Capex $\times$ 10 years
	$= \$500,000$
Depreciation per kg_{GN} from NG	$= \frac{\$500,000 \text{ annual cost}}{350,000\ kg_{GN} \text{ produced}}$
	$= \$1.43$ per kg_{GN-NG}
Annual depreciation of CNT biogas plant	$= \$15,000,000$ Capex $\times$ 10 years
	$= \$1,500,000$
Depreciation per kg_{CNT} from biogas	$= \frac{\$1,500,000 \text{ annual cost}}{350,000\ kg_{GN} \text{ produced}}$
	$= \$4.29$ per $kg_{CNT-biogas}$
Annual depreciation of CNT – syngas	$= \$16,000,000$ Capex $\times$ 10 years
	$= \$1,600,000$
Depreciation per kg_{CNT} from syngas	$= \frac{\$1,600,000 \text{ annual cost}}{350,000\ kg_{GN} \text{ produced}}$
	$= \$4.57$ per $kg_{CNT-syngas}$
Annual depreciation of CNT NG plant	$= \$12,000,000$ Capex $\times$ 10 years
	$= \$1,200,000$
Depreciation per kg_{CNT} from NG	$= \frac{\$1,200,000 \text{ annual cost}}{350,000\ kg_{GN} \text{ produced}}$
	$= \$3.43$ per kg_{CNT-NG}
Annual depreciation of GO	$= \$4,000,000$ Capex $\times$ 10 years
	$= \$400,000$
Depreciation per kg_{GO}	$= \frac{\$400,000 \text{ annual cost}}{350,000\ kg_{GN} \text{ produced}}$
	$= \$1.14$ per kg_{GO}

CHAPTER SIX

Functionalization and In Situ Polymerization

6.1 FUNCTIONALIZATION

Within our discussion domain, functionalization refers to the process of changing the high-energy states of the dangling bonds to groups that provide useful characteristics. In many applications, the graphitic nanofibers (GN) or any other graphitic material must have some functional groups attached to be effective for a wider range of applications. For example, $-COOH$ and $-OH$ functionalities make the materials hydrophilic and provide ion exchange sites for electrophilic reactions. In the following referenced work the material used in obtaining the experimental data has mostly been referred to as graphene or graphene oxide. I cite them here and in the discussion to assert the ability of GN to perform equally or better. Note that in all the reported work, functionality is always provided by chemical groups attached to the edge zones of the graphene platelets, a characteristic natural to GN. The moieties possible on these edges are not always mentioned, so comparison with other commercial materials that are already commercially available with these immobilized functional groups is not easy. However, the examples covered by the research findings give us an opportunity to draw parallels to GN, and give insight on the versatility of GN. The graphene or GO is prepared from Flake graphite as the starting material, with all the disadvantages I have explicated before. Though GO is called graphene oxide, you will notice mostly it is graphite oxide that is being referred to by the popular name. GN are superior to GO materials that are prepared from flake graphite using Hummers' (or modified Hummers') method. Even if fewer layer fibers are desired, GN offer much lower overall costs and predictable capacities of the immobilized functional groups.

Functionalization of GN requires only a mild acid wash. The edge sites form $-COOH$ and $-OH$ groups very readily. Flake Graphite functionalization can only take place after harsh oxidation has created active

Graphitic Nanofibers.
DOI: http://dx.doi.org/10.1016/B978-0-323-51104-9.00006-6

sites on the basal planes. The –COOH groups act as active weak acid ion exchange sites and provide many benefits in applications discussed later. A dilute acetic acid wash converts substantially all other groups to –COOH groups giving a very consistent functionality profile. You will notice that most of the functionalization examples I have cited have the –COOH group as the precursor to the final product.

6.1.1 Metals incorporated

Carboxylic groups have a strong affinity for heavy metals. Metal ligands can be attached to the edge groups to perform various further functions. Carboxylic acid functionalized graphene oxide–copper (II) sulfide nanoparticle composite (GO-COOH-CuS) was prepared from carboxylated graphene oxide and copper precursor in dimethyl sulfoxide at room temperature. The as-synthesized GO-COOH-CuS nanocomposite exhibited excellent photocatalytic degradation performance of phenol and rhodamine B (surrogate for dyes), high antibacterial activity toward *Escherichia coli* and *Bacillus subtilis*, and good recovery and reusability. Costs permitting, GO-COOH-CuS has potential for water treatment by photocatalytic reactions [202]. TiO_2 is the current catalyst of choice for photocatalytic reactions.

6.1.2 Chelates/amines

Nitrogen groups react via an electrophilic exchange with the edge groups. Ethylene diamine group on GO acted as a chelating agent [203] to remove heavy metals from aqueous streams, though the extensive comparison in this work exercise does not mention the most popular media in water treatment today for immobilized chelating agents, i.e., imminodiacetic functionalized ion exchange resins. Herringbone GN (finally! the right nomenclature.) were surface-derivatized with reactive linker molecules derived from four diamines and three triamines. Surface sites of as-prepared GN were oxidized to carboxylic acid groups by nitric acid and covalently bound to seven different linker molecules containing pendant amino groups using carboxylate amidation chemistry [204].

6.1.3 Sulfonation

Liu et al. [205] developed sulfated graphene oxide ($GO - OSO_3H$) with $-OSO_3H$ groups attached to the carbon basal plane of reduced GO surrounded with edge-functionalized –COOH groups. The resultant $GO - OSO_3H$ exhibited excellent hole extraction layer characteristics for polymer solar cells (PSCs). The reduced basal plane helped to increase conductivity while its $-OSO_3H$/$-COOH$ groups enhanced solubility for solution processing. The improved conductivity compared to GO (1.3 S/m vs 0.004 S/m) lead to a power conversion efficiency (4.37% vs 3.34%) of the resulting PSC devices.

One can also extend the procedure in this work for a more realistic application in water treatment and homogeneous catalysis. Sulfonic groups act as strong acids and are also the immobilized functional group on ion exchange resins that remove cations from salt solutions. By contrast, $-COOH$ groups have a pK_a of ~ 4.2 and can only remove cations associated with alkalinity, (with the exception of heavy metals). Sulfuric acid is also used as a homogeneous catalyst in the widely used esterification reaction in industry. The location of ptotonated sulfonated groups on the edge sites of GN could provide immobilized Lewis acid sites for carrying out homogeneous catalytic reactions with nanomaterial heterogeneous methods. Centrifugation would allow the functionalized GN to be recovered and reused. In another application, Neelakandan et al. [206] have fabricated proton exchange membranes by sulfonated poly(1,4-phenylene ether ether sulfone)/poly(ether imide)/sulfonated graphene oxide. Sulfonated graphene has also been used to improve the conductivity of conductive polymers such as polyaniline [207].

Sometimes, functionalization is also the term used for in situ polymerization using GO. The procedure usually requires the GN or oxidized GN to be functionalized before the reaction can take place in the homogeneous phase. We will look at these in the next section.

6.2 IN SITU POLYMERIZATION

Besides physical dosing of carbonaceous nanomaterials into polymers by shearing forces or hydrogen bonding (discussed later), we can drastically reduce the dose of the nanomaterials, by incorporating the nanomaterials into the polymer backbone. This is achieved by having them be present during polymerization. In situ polymerization is the term used to refer to polymerization within a mixture. This process is not commonly used in industry, presumably due to cost. Consider the Ziegler–Natta reaction: the dominant process in industrial production of polymers of alpha olefins (1-alkenes) such as polyethylene, polypropylene, polybutadiene, polyisoprene, etc. Ziegler–Natta reactions are carried out by a heterogeneous catalyst on $MgCl_2$ supports in a fluidized bed. For the enhancement of polymer properties, in situ polymerization would be an attractive route in some of these processes and are worth a closer look.

6.2.1 Homogeneous Ziegler–Natta reaction

The homogeneous version Ziegler–Natta reactions offer some possibilities for the incorporation of GN into the polymer matrices. The catalysts used for homogeneous catalysis are metallocenes, which are also known as "sandwich compounds" and have a general composition of Cp_2MCl_2 (where Cp is an organic ligand and M is a metal such as titanium or zirconium). There is also a cocatalyst, typically methyl aluminoxane (MAO) with the structure $-[O - Al(CH_3)]_n-$. The organic ligands Cp are usually anions of cyclopentadiene derivatives.

6.2.2 GN as catalyst carriers or cocatalyst carriers

GN can be used as catalyst carriers, or bear the immobilized cocatalyst (e.g., MAO) for these reactions. The polymerization would occur in situ and incorporate the GN within the polymer structure, which is a highly desired consequence in the quest for composites. This route to enhancement by GN eliminates dispersion issues faced with melt blending or shear-type mixing. MAO is a Lewis acid and the –COOH functionality of GN would strongly attract MAO due to the aluminum. The polyethylene and polypropylene composites of GN can be made in a manner similar to that described with 'graphene'. [208,209] (see Figs. 6.1–6.3). Polypropylene/graphite composite has been prepared by a pseudo homogeneous Ziegler–Natta reaction by attaching a Grignard catalyst to GO and mixed with the $TiCl_4$, which is the normally used heterogeneous catalyst [210] *(If graphite was replaced by GN, a higher number of equivalents per gram would be embedded).*

6.2.3 Nonaqueous solvents

In general, functionalization can be achieved either by (1) covalent [194,211–213] or (2) noncovalent means [214,215].

Covalent bonding is stronger but may marginalize applications that require an improvement in conductivity because electron transport may get affected due to conversion of carbon from sp^2 to sp^3 during covalent bonding. Due to the high –COOH and –OH moieties on oxidized GN, organic groups can be attached with ease. Porphyrins,

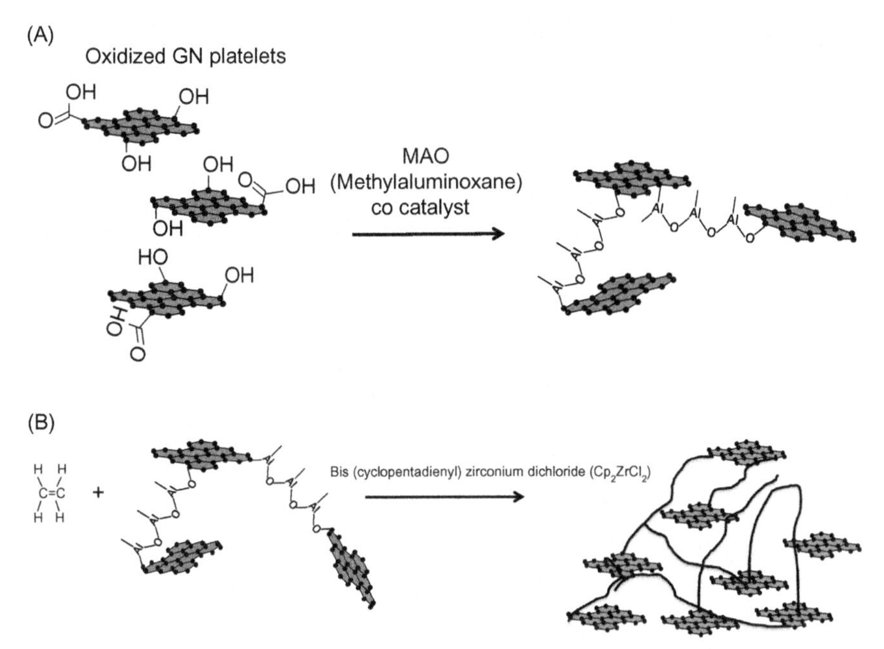

Figure 6.1 Ziegler–Natta in situ polymerization of polyethylene [208]. (A) Functionalizing with MAO. (B) Catalyzed in situ polymerization with GN embedded in the polymer matrix. *GN*, graphitic nanofiber; *MAO*, methyl aluminoxane. *NOTE: GN is used in the illustration as a substitute for Graphite used in the referenced work.*

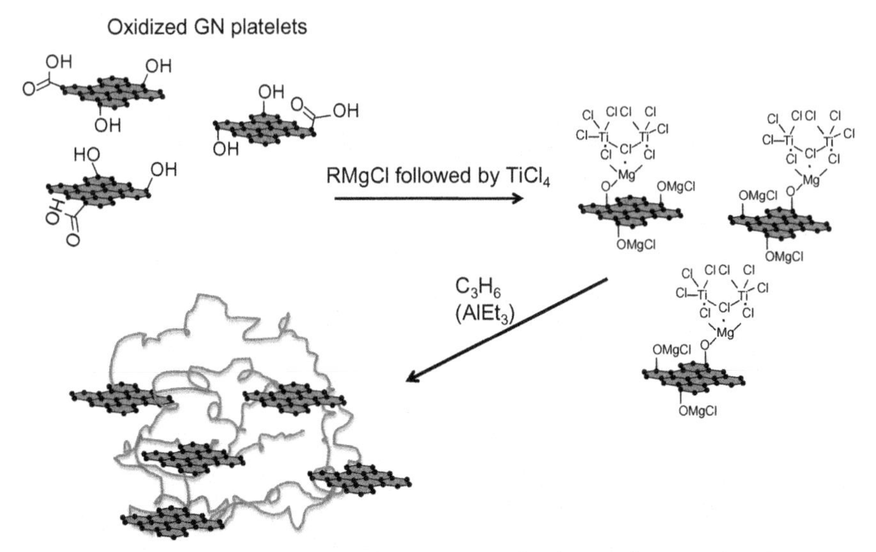

Figure 6.2 Ziegler–Natta in situ polymerization of polypropylene [210]. *NOTE: GN is used in the illustration as a substitute for Graphite used in the referenced work.*

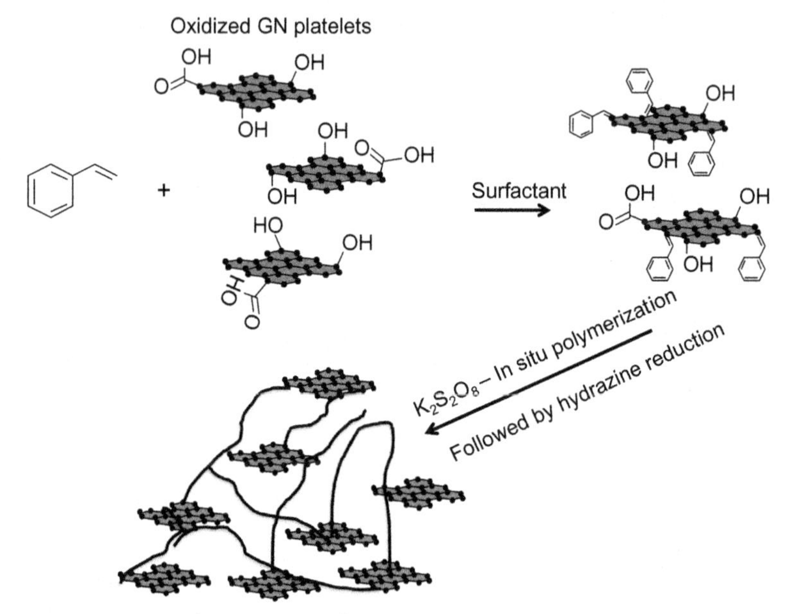

Figure 6.3 In situ polymerization of styrene. Note: Diagram depicts use of GN as opposed to GO used by authors in the referenced work.

phthalocyanines, and azobenzene have been covalently attached on the GO nanoparticles to enhance optical properties [216–219]. *Graphite nanoparticles could be replaced by GN, providing a much higher surface area, consistent shapes, and predictable capacity for the organic groups.*

For non-aqueous solvents, GN would need to be functionalized for miscibility in the solvent and interaction with the monomers. Organophilic graphitic nanosheets are usually prepared by reacting GO with octadecylamine (ODA) followed by a reduction process. If refluxed long enough, the ODA chain will eventually reduce the graphite, without any additional steps [220]. Polystyrene (PS) and poly (methyl methacrylate) (PMMA) blends with GO/ODA resulted in conductive composites with lower percolation thresholds. *GN is an excellent substitute for graphite here.* To get a stable dispersion of oxidized GN into the polymer, the GN would have to be functionalized with a group that would produce stearic hindrance due to its size, and by default, have higher exposure between the GN and the matrix.

6.2.3.1 Composite GN—polyurethane

Similar procedures can be deployed for polyurethane composites. In situ polymerization of 2% by weight GNS. (Not GN, but Graphene Nano

Sheets – (I missed this superfluous surname in my earlier descriptions) and polyurethane produced a composite with double the tensile strength and modulus of the base material. The nanocomposites also displayed high electrical conductivity and thermal stability [221]. GN could perform just as well, and possibly double the values of storage modulus and tensile strength of polyurethane (PU). PU composites reinforced with GO and r-GO were prepared by in situ polymerization (see Fig. 6.4). Characterization of oxidized, then reduced graphite showed that, due to the formation of chemical bonds, the graphite was dispersed well in the PU

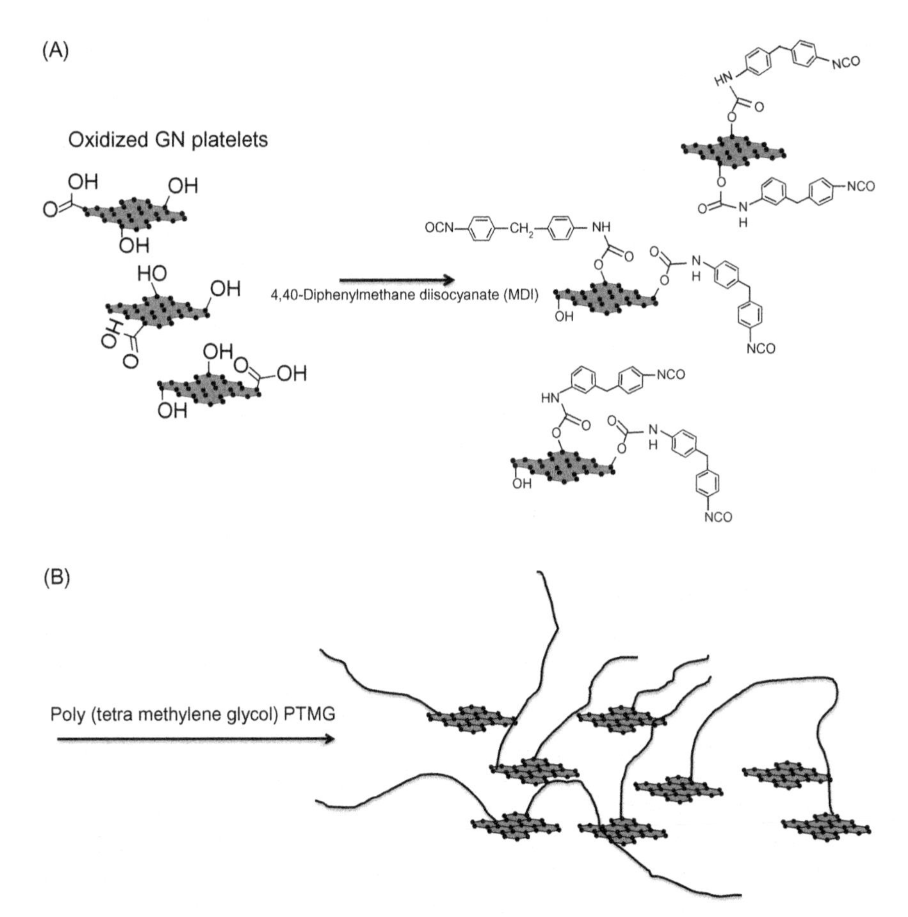

Figure 6.4 In situ polycondensation of MDI and PTMG for polyurethane in the presence of graphite sheets. (A) Embed isocyanate onto graphene platelets. (B) Polymerization. *MDI*, 4,40-Diphenylmethane diisocyanate; *PTMG*, poly(tetramethyleneglycol). NOTE: GN is used in the illustration as a substitute for Graphite used in the referenced work. *Adapted from X. Wang, Y. Hu, L. Song, H. Yang, W. Xing, H. Lu, In situ polymerization of graphene nanosheets and polyurethane with enhanced mechanical and thermal properties, J. Mater. Chem., 21 (2011) 4222–4227.*

matrix. In situ polycondensation of 4,40-Diphenylmethane diisocyanate (MDI) and poly(tetramethyleneglycol) (PTMG) in the presence of graphite was performed. (Back of napkin cost analysis - 1) as tested, cost addition ~$3.52 per kg of PU. If substituted by GN, *$1.02 per kg of PU.* Refer to Table 5.1 for costs. *Current cost of polyurethane ~ $1.70 per kg.* Enhancement by GNS triples the cost of PU, while enhancement by GN adds 60% to the original cost of PU. *Is either one worth it? Will industry embrace it with open arms?*

6.2.3.2 Composite GN—poly (methyl methacrylate)

Composites of GN and PMMA may be prepared by in situ polymerization. Pramoda et al report 'graphene' enhanced nanocomposites [222].

GO with –COOH groups at the edges was functionalized with ODA. The functionalized GO was reacted with methacryloyl chloride to incorporate polymerizable –C=C–platelet edges and planes, and in situ polymerization of methylmethacrylate with covalently bonded PMMA–graphene nanocomposites were obtained. The nanocomposites showed significant enhancement in thermal and mechanical properties compared with neat PMMA. An important parameter T_g increased from 119°C to 131°C, with just 0.5 wt% graphite, accompanied by a storage modulus increase from 1.29 to 2 GPa. *(Back of napkin cost analysis: PMMA commodity price: ~ $2.0 per Kg. As tested material ~ $0.90 cost addition per Kg – If substituted by GN ~ $0.26 cost addition per Kg. Representing a cost increase of 45% and 13% respectively. This is a clear case where GN is a very probable go and GO a definite no.)*

6.2.3.3 Composite GN—polyimide

Comparable Polyimide-GN composite films can be prepared with reference to technicques shown by Luong et al using 'graphene' sheets [223]. Polyimide/GN composite film can be prepared by in situ polymerization [223].

GO was functionalized with ethyl isocyanate and dispersed in *N,N'* dimethylformamide. The characterization of the graphitic functionalized product was not described clearly by the authors. They assume the thickness of a single graphene sheet is 0.34 nM, which is usually the *d* spacing in graphite. Then they report 1-2 nM thick flakes as "single" flakes explaining the extra dimension as a result of basal presence of oxidation products. The dispersion was then used as a media for the situ polymerization, and films of 3μ were fabricated. Using Kapton® (Dupont) as an example, approximately 13 square meters of film with the thickness used by the authors would weigh 1 Kg. A 0.38 wt% of the functionalized graphite resulted in a Young's modulus increase of 30% compared to pure PI film. The corresponding tensile strength increased from 122 to 131 MPa. In addition, the electrical conductivity was increased by more than eight orders of magnitude. The cost to benefit ratio of this application is very attractive. 1 Kg of Kapton® can be estimated at ~$1000. 0.38% of GO would cost ~$0.67, (Refer to Table 5.1) with the undesired presence of oxidation products on the basal plane. The application involving films beg for higher tensile strength, allowing much thinner films, and higher conductivity for anti static properties. Even with GO at costs an order of magnitude higher, the application is very attractive, based on the data from the authors. In this case, GN would have a technical advantage of discrete edge-only sites. The cost difference is inconsequential.

6.2.3.4 Grafting

Within in situ polymerization, there is a defined procedure called Grafting. Grafting is defined in two manners: (1) to and (2) from. Generally in grafting "to" method the polymer chains are synthesized first, and then attached with the functional groups of graphite oxide or graphite (reduced or as produced) with an aromatic surface. The direct covalent link of the functional polymers on the oxidized graphite surface using esterification and amidation are examples of this method. I will discuss the published materials on this method.

Esterification: The esterification reaction between the carboxylic acid group of oxidized graphite and the hydroxyl groups of poly (vinyl alcohol) in the presence of N,N_0 dicyclohexylcarbodiimide (DCC) and 4-dimethylaminopyridine (DMAP) produces a strong elastic thin film that shows a dramatic shifting of T_g from 70°C to 90°C indicating that the oxidized graphite nanosheets reduce the polymer chain mobility substantially. The crystalline PVA is transformed into amorphous material and the degradation temperature increases by 100°C [224,227] *Not enough information given for a cost analysis.*

Biocompatible and water-soluble polysaccharides (e.g. hydroxypropyl cellulose and chitosan (LMC)) films can also be combined with GN via esterification reaction, shown via "graphene" by Yang et al [225].

An interesting note: Chitin is present in crustaceous marine animals and has a strong affinity for oil (fat). Presumably while still on the shell, it attracts fat from the animal's body toward the shell to enable the shell to come off easily when the time comes for shedding it. Chitin is abundantly available as a waste product from the processing of crustaceous seafood. When cross-linked, it can be converted to solid beads (chitosan). It has a high level of quaternary amine groups and other nitrogen groups that can be used for many purposes, including the use as an adsorbent for undesirable anionic species in water, such as nitrates, fluorides, chromates, etc. Due to its fat absorption qualities, it is also used in wastewater treatment to remove oils and greases from the wastewater of many food industries, such as chicken processing plants. Lately, it has been a popular addition on the nutritional supplement shelves with the claims it "ties up" the fat in the food you eat, so you can satisfy your cravings and vanity at the same time. Chitosan is insoluble in water. To make it water soluble for so many applications, it is usually cross-linked with glutaraldehyde. Chitosan can also be easily cross-linked to other polymers to form hydrogels. Cross-linking occurs through a nucleophilic attack on the carbon of the cross-linking agent by the nitrogen in the amino group. These polymers have some physical limitations for widespread use, especially chitosan beads. It is a food grade material that could be very useful if it was made conductive or mechanically strong. If aldehyde groups can be used to functionalize oxidized GN for cross-linking chitosan, the stronger versions would have a significantly large market demand. Glutaraldehyde can also be used to cross-link oxidized graphitic nanoparticles (such as GN) to give a continuous paper or membrane structure [226].

The esterification reaction between GN and poly(vinyl chloride) (PVC) can be achieved introducing a hydroxyl group on the PVC nucleophilic substitution. PVC modified in this manner showed a 30°C increase of T_g for the 1.2 wt% GO content indicating that oxidized GN could also retard the mobility of the polymer chains and exhibit a mechanical reinforcement causing up to a 70% increase of the storage modulus. [227,228] (Fig 6.5). *(Back of napkin cost analysis: PVC commodity price: ~$4.0 per Kg. As tested material ~$2.11 cost addition per Kg – If substituted by GN ~$0.61 cost addition per Kg. Representing a cost increase of 53% and 15% cost increases respectively. Again, GN is a very probable go and GO a definite no.)*

Water-soluble GO has been grafted with poly(*N*-isopropylacrylamide)(PNIPAm) [229]. Alkyne derivative of GO via amide linkage was prepared and then coupled with PNIPAm containing azide end group. This functionalized GO lowered critical solution

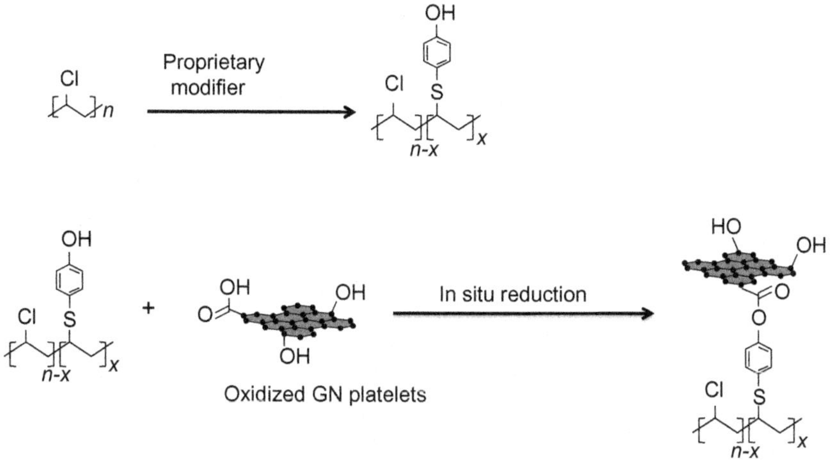

Figure 6.5 PVC enhancement in situ. NOTE: Diagram depicts use of GN as opposed to GO used by authors in the referenced work.

temperature (LCST) to 33°C from the normal hydrogel LCST of 37.8°C. This LCST improvement is significant for drug delivery applications. Similarly the azide-terminated poly(ethylene glycol) (PEG) [230] and azide-terminated PS are grafted with alkyne derivative of graphene [231]. The azide-terminated PS, PMMA, poly(4-vinyl pyridine) (P4VP), and poly(dimethyl aminoethyl methacrylate) (PDMA) polymers have been synthesized by RAFT polymerization using azide-terminated chain transfer agents and they are subsequently grafted onto the graphene surface [232,233]. Hydrogels with a lower LCST can be beneficial for drug delivery and other high value applications. The cost of GN for these applications would be quite palatable.

GN can also be used for condensation reactions. The condensation reaction between the carboxyl acid groups of oxidized GN with active amino groups of nylon chains can produce a high strength composite, similar to work reported by Xu et al using 'graphene' [234] (Fig. 6.6).

Sun et al. have synthesized GO hydrogel by direct covalent linking with poly (*N*-isopropyl acryl amide-*co*-acrylic acid) (PNIPAm-*co*-AA) in water [235]. The hydrogel shows dual thermal and pH response with good reversibility. Our lab has also successfully grafted oxidized GN into reversible hydrogels to make a dehumidification apparatus. Upon reaching capacity, the hydrogel is reversed and dewatered by a slight electrical perturbation through the conductive material [236].

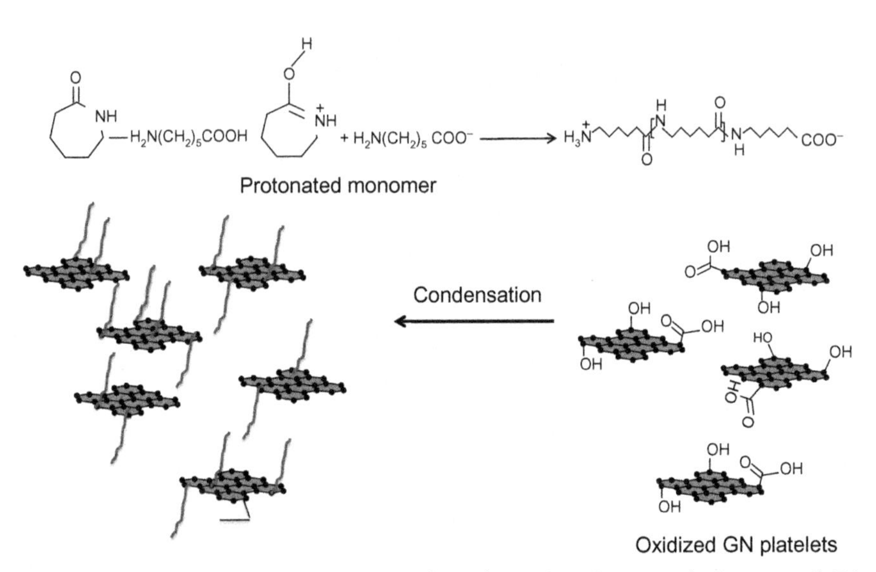

Figure 6.6 Condensation reaction in situ for nylon. NOTE: Diagram depicts use of GN as opposed to GO used by authors in the referenced work. *Adapted from Z. Xu, C. Gao, In situ polymerization approach to graphene-reinforced nylon-6 composites. Macromolecules 43 (2010) 67.*

The noncovalent functionalization on graphitic surfaces with small molecules and polymers has been utilized widely [237–243]. In the noncovalent interactions, H-bonding and π–π stacking on the graphene surface play an important role. The noncovalent functionalization possesses significant advantages over covalent functionalization as it enhances the solubility without the alteration of extended p conjugation of the graphene sheet, whereas in the covalent functionalization sp^3 defects are created on the graphene ring.

It is unclear if in situ polymerization or grafting will be adopted as a route in the manufacturing of polymers. The opportunity to enhance properties should be enticing enough, provided the overall costs for the enhanced materials can be within acceptable ranges. For sure, the role of GN in very specialized applications can be significant if used as a medium. When reaction equilibrium is far to the right, and the desired product is an intermediate product, say after the desired level of hydrogenation, oxidation, etc. some clever techniques can be applied, using oxidized GN as a suspended catalyst support. Highly selective hydrogenation of Resorcinol to 1,3-cyclohexanedione through π-conjugate interaction has been achieved in one such technique [244].

I have deliberately used the word graphite many times in the above discussion, because the contributing authors' descriptions of the materials in the experimental sections define either oxidized graphite or functionalized graphite. The titles of the research papers use the term "graphene" and "graphene oxide" when the experimental work involved does not mention what I define as graphene or graphene oxide.

CHAPTER SEVEN

Applications

7.1 POLYMER ADDITIVES

On a macroscopic scale, physical properties of polymer materials are a function of their phase state. In reality, the phase state is often a complex and concealed function of the structure and molecular interactions. Adding copolymers and other materials modifies these characteristics. The potential advantage of nanomaterial additives is that their small size can produce improved mechanical properties without degrading the energy absorption (impact) properties of the composites, a disadvantage shared by most conventional macro composites [245].

Size matters. Piggott [246] concluded that nanoparticles are more effective reinforcements than their conventional counterparts because a smaller amount of nanoparticles could lead to a larger improvement in the mechanical properties of the polymer matrix. In addition, nanoparticles serve as a more efficient stress transfer medium to transfer the stress from the matrix to the reinforcements due to the increased surface area and good adhesion at the interface.

7.1.1 Improving mechanical strength

Mechanical strength is one of the two properties most investigated in the polymer enhancement applications. Carbon fibers and microfibers have been used in the polymer industry due to their higher tensile modulus, strength, and excellent electrical and thermal properties. Applications have mostly been high end such as aerospace (Boing 777 is mainly made of composites), automobiles, sporting goods and military. In recent years, polymeric *nano* composites have attracted research interest both in industry and in academia, which represent a radical alternative to conventional filled polymers or polymer blends [247]. The studies used to center around CNTs, but now they seem to be focused on Graphene.

Most carbon nanofibers have round cylindrical surfaces. However, many of the references cited actually used graphitic nanofibers, but termed them as carbon fibers or CF. I have distinguished between the

Graphitic Nanofibers.
DOI: http://dx.doi.org/10.1016/B978-0-323-51104-9.00007-8

two by properly naming the graphitic fibers as GN. The cylindrical carbon fibers typically have evenly distributed surface electron density, similar to CNTs, but CNF do not have concentric walls or hollow centers. CF diameters range from 100–200 nM diameters. Surface van der Waal (VdW) forces dominate. If during formation, the distance between fibers is less than 0.4 nM, VdW forces create a very strong interaction between the fibers, causing entanglement. Of the limited amount of CNT/polymer composites being produced, most well dispersed, electrically conductive and thermally stable composite materials employ high shear methods like twin-screw extrusion to break this entanglement [248,249].

High shear methods are energy intensive and require a change in the production process with large capital investments. Though efficient at breaking the VdW forces to make a well dispersed composite, high shear methods also reduce the aspect ratio of the nanofibers by literally breaking the fibers. This further reduces the active surface area, resulting in the failure to realize the full potential benefits for polymer enhancement.

Graphitic nanofibers (GN) are entirely made up of sharp graphene plane edges, and mostly held together with sp^2 bonded electrons, with the VdW forces holding the graphene planes together in different directions relative to the fiber axis. If produced with high purity, GN therefore exhibit less entanglement behavior when dispersed in solvent suspensions. Dispersion in the polymer matrix of a graphitic product with edges to repel other fibers does not require high shear methods. The highly active edges (if oxidized with mild acid treatment) are capable of forming π bonds and hydrogen bonds with the base polymer. It follows that much better performance can be realized if the nanofibers are not merely suspensions of nanomaterials in the melt batch, forced there by the high shear process, but actually have 1:1 interaction with the polymer, giving the polymer "backbone" the support and or characteristics required for higher strength.

Attempts have been made by others to overcome agglomeration with complex multistep non-high shear procedures [250]. The unique characteristic of GN can be utilized to form polymer composites by having sites on the base polymer interact with the GN.

One such favorable interaction is *hydrogen bonding*, which has been reported by many as a mechanism for making multi-polymer blends [251–255].

Polymers containing ternary amide groups (see Fig. 7.1), such as poly (*N*-vinyl pyrrolidone) (PVP) are potentially good proton acceptors because of the basicity of the functional groups [256]. PVP is reported to form

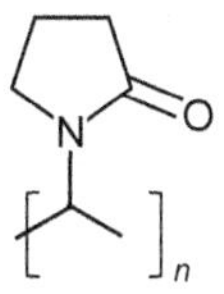

Figure 7.1 Poly(*N*-vinyl pyrrolidone).

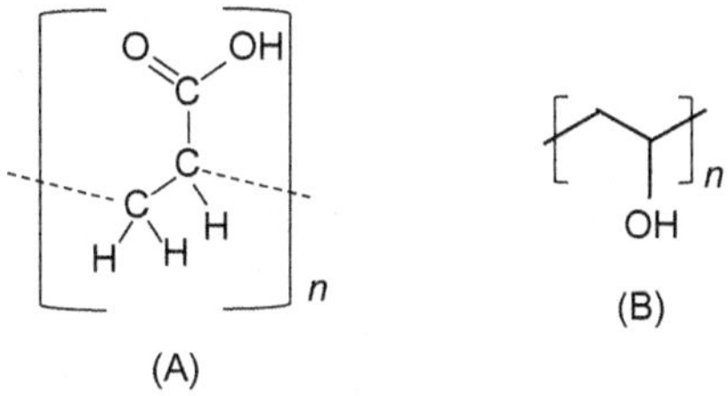

Figure 7.2 (A) Polyacrylic acid and (B) poly vinyl alcohol.

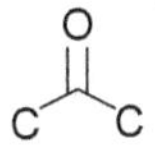

Figure 7.3 C=O.

strong H-bonded complexes with polymers bearing proton donating functional groups in the repeat units of their backbones, such as carboxyl groups in polyacrylic and polymethacrylic acids (PAA, PMA). See Fig. 7.2 [257,258] and hydroxyl groups in poly(vinyl alcohol) (PVA) [259].

The *hydrogen bond* is a strong intermolecular force influencing neutral (uncharged) molecules. Hydrogen forms polar covalent bonds to more electronegative atoms such as oxygen (Fig. 7.3), and because a hydrogen atom is quite small, the positive end of the bond dipole (the hydrogen) can approach neighboring nucleophilic or basic sites more closely than can other polar bonds. Since organic compounds have C–H bonds, a useful rule is that the observed Fourier Transform Infra-Red (FTIR) absorption in the 2850–3000 cm^{-1} is due to sp^3 C–H stretching, whereas absorption above 3000/cm is from sp^2 C–H stretching and sp C–H stretching if it is near 3300/cm. Molecules having both hydrogen bonding donors and acceptors located such that intra molecular hydrogen bonding is favored, display slightly broadened O–H stretching absorption in the 3500–3600/cm range.

The oxygen atom in the C=O bond contains two lone pairs of electrons :O: in hybrid orbitals, which are oriented at 120 degrees to each other. During the complex formation, the donor hydrogen aligns itself with one of the lone pairs to form a bond like O–H···O=C to indicate that the H bond joins together with two bonds and not with two atoms. This position is most favorable for the maximum interaction to occur between the lone pair atomic dipole and the O–H bond. Resulting in the formation of a 1:1 complex. When the OH bond vibrates, the lone pair electron also vibrates in consonance with the O–H bond and thus contributes to the changes in the dipole moment. In the case of C=O frequencies typically observed, the C=O induces a moment in the O–H bond, and this would lead to an increase in the intensity of the carbonyl stretch. Because C=O has low polarizability, the induced moment change will be small compared to the change in the O–H pole dipole.

In our laboratory, we titrated oxidized GN (oxidized by mild acid treatment) and established a high molar presence of –COOH groups (5.32×10^{-4} moles/g_{GN}) and oxygen concentration (34.2%) with (third party) X-ray photoelectron spectroscopy (XPS). Group functionality was confirmed by FTIR absorption bands (2880–3360/cm) and (3660/cm) indicating a highly functionalized surface rich in –COOH and –OH groups. Oxidized forms of the GN are therefore well suited for matrix interaction with polymer backbones having basic groups and hydrogen-bonding possibilities. Typically, H-bonding groups would be located in monomer units of backbones. One must account for the need to balance the gain in enthalpy and the loss in entropy. Therefore, most likely, the first hydrogen bond appearing between the chains of complementary polymers would initiate the process of cooperative binding, which resembles a fast zip formation and leads to a ladder-like structure between platelet type GN and the polymer.

Relating these activities to polymers, the free energy change for the hydrogen bond formation of oxidized GN in methacrylates is expected to be in the order methyl < ethyl < butyl. This is due to the difference in basicity of the alkyl methacrylate groups. The basicity is attributed [259,260] to the negative inductive effect of the alkyl groups, which, in turn, increase in the order of methyl to butyl, where the electron contribution of the butyl group to the C=O group is significantly greater than that of the methyl group. Therefore, the strength of the intermolecular hydrogen bond formed between the C=O group and an oxidized GN's proton is dependent on the basicity of the C=O group, the acidity of the GN–OH proton, and the intermolecular distance between these acid and base sites. We can conclude that a high surface area material with high degree of acidity would form stronger hydrogen bonds with the polymer backbone.

The free O–H band intensity increases with increasing GN concentration but at the same time, the reverse trend is expected for the carbonyl absorption band. This would point to the formation of 1:1 complex between the free hydroxyl and carbonyl groups [261].

For maximum effects, Whetsel and Kagarise [262] and Pillai et al. [263] have done extensive work to study the effect of different types of solvents with strong interactions and weak interactions on the carbonyl absorption bands which are representative of compounds of esters, ketones and aldehydes. They confirmed the existence of the formation of 1:1 and 1:2 complexes in the different systems studied. Acrylic esters (methyl acrylate, ethyl methacrylate, and butyl methacrylate (BA)) are the important industrial chemicals they studied.

As polymer additives, GN are far more attractive than their *carbon* counterparts for enhancing mechanical qualities such as a better modulus, better thermal degradation properties, etc. I emphasize one cannot assume simply graphitizing carbon nanofibers can provide the qualities of synthesized graphitic materials. Graphitizing cylindrical carbon fibers recrystallizes their surfaces into discontinuous crystallites. This discontinuous structure, as expected, improves the crystallinity of the carbon but does not give optimum mechanical or electrical properties to composites in which they are used. The structure will not even impart the lowest resistivity to the fiber itself. This is because the short, nested conical crystallites interface with grain boundaries that lower the fiber's longitudinal mechanical strength, stiffness, and electrical conductivity [264].

These oxidizing treatments can add oxygen atoms to 1/4 of the fiber's surface [265]. Lakshminarayanan et al. [266] studied vapor grown carbon nanobifers (VGCNF), which had been oxidized with nitric acid. Shallow micropores were created by acid treatment, which covered the fiber surface and made the fibers dispersible in water, with XPS measurements showing up to 22% surface oxygen coverage.

In summary, making optimal composites of fibers requires adequate fiber-matrix adhesion, which is governed by the chemical as well as physical interactions occurring at the fiber—matrix interface. If the fiber—matrix adhesion is poor, the composite may fail at the interface, reducing in particular the tensile strength of the composite. While extensive literature describing the surface treatment of conventional carbon fibers by such methods as oxidation in the gas and liquid phases or anodic etching exists, practical work is still rudimentary for carbon nanofibers.

In polypropylene composites, it appears that modest surface oxidations of GN, up to 4% surface oxygen atoms, makes composites with optimum tensile strengths [265]. Further increases in oxygen concentration appears to decrease composite tensile strength. This behavior is attributed to the less polar behavior of the low energy surface of polypropylene, which may have difficulty bonding to a highly oxygenated nanofiber surface. For such applications, GN would not require any treatment beyond the mild acid wash used for catalyst recovery.

On the other hand epoxy composites give improved tensile properties with the nanofiber surfaces strongly oxidized. Lafdi and Matzuk [267] reported a 35% strength improvement and a 140% modulus improvement with 4 wt% loading of highly oxidized VGCNF. Other ways of oxidizing the fibers during production have been reported. Patent by Glasgow and Lake describes a scheme utilizing the addition of oxidizing gases, preferably the mild oxidizer CO_2, to increase both the surface area and surface energy of VGCNF, but with possible deactivation of the vital growth catalysts [268]. None of these exotic methodologies are required in the synthesis of GN. The required characteristics can be produced by proper operating conditions of the reactor itself *(cost if GN was dosed in epoxy: $2.04 per Kg. This would be virtually impossible to sell to the low end epoxy industry, but could be justified for high end applications.)*.

7.1.2 Improving electrical characteristics

Electrons flow in the lateral direction of graphene planes. Ribbon-type GN having graphene planes that are parallel to the fiber axis will exhibit high conductivity functionally similar to SWCNTs. Most of the work on enhancement of electrical properties of polymers has been done with CNTs or Graphene. Ribbon GN enhanced thermoplastic composites can offer permanent conductivity at electrostatic discharge levels with very low percolation threshold values (percent of nanofibers by weight of total polymer where the composite becomes conductive) without affecting processing and other beneficial properties of the host polymer. This ability makes these materials of particular interest for applications in electrostatically painted automotive parts, ESD plastics for sensitive electronic packaging and manufacturing, ESD carpets in lighter colors (no more sparks from the fingertips), etc.

Agglomeration of fibers has to be addressed in this application. It might be argued that agglomeration may help conductivity, but it would also translate to larger percolation thresholds, because as common sense would dictate, a continuous fiber network has to be established for a polymer to become conductive.

Much of the technical comparison between GN characteristics and those of commercially available nanofibers applies when the objective is to make the polymer electrically conductive.

- Graphitic materials will outperform carbon materials.
- Matrix/nanofibers bonding and the resultant "ladder" effect will ensure very consistent dispersion and continuity of the fibers.
- Ribbon-type GN can also be embedded on the surface of the polymer during extrusion to provide surface conductivity, severely reducing the charge holding capacity of the polymer in ESD applications.

Electrical conductivity improvement can also be achieved by addition of non carbonaceous materials. I will stick to our focus and not discuss those composites. I will do a cost analysis on a case by case basis, whenever sufficient data is given. In the interest of brievity, and the purpose of this book, I have completely omitted applications I studied which were not even close to economic justification.

7.1.3 Polymers

7.1.3.1 Polycarbonate

Polycarbonates are one of the fastest growing engineering plastics in the world, as new applications are increasingly being devised. This is one of the polymers I feel can succeed as a composite with GN, costs permitting. Polycarbonates are one of the most widely used engineering thermoplastics due to their physical characteristics such as high thermal and electrical stability, excellent toughness and stiffness, biological inertness, cost effectiveness, ready recyclability, and high resistance to fracture and impact. Bullet or shatter-resistant glasses, lenses and windows, flame-retardants, automotive, laptop and cellphone parts, and compact discs are all typical applications for polycarbonates. The demand for polycarbonates is especially expected to rise in the automotive industry in the coming years, as the industry is constantly in search of substances that will reduce the final weight of a vehicle, a factor that obviously affects fuel efficiency.

GN (cupped) have been successfully blended with polycarbonate for electrostatic dissipation (ESD) applications [269] and mechanical property enhancements [270]. Thermally exfoliated and oxidized graphite (GO) has also been evaluated for electrical and mechanical properties [271] (Fig. 7.4).

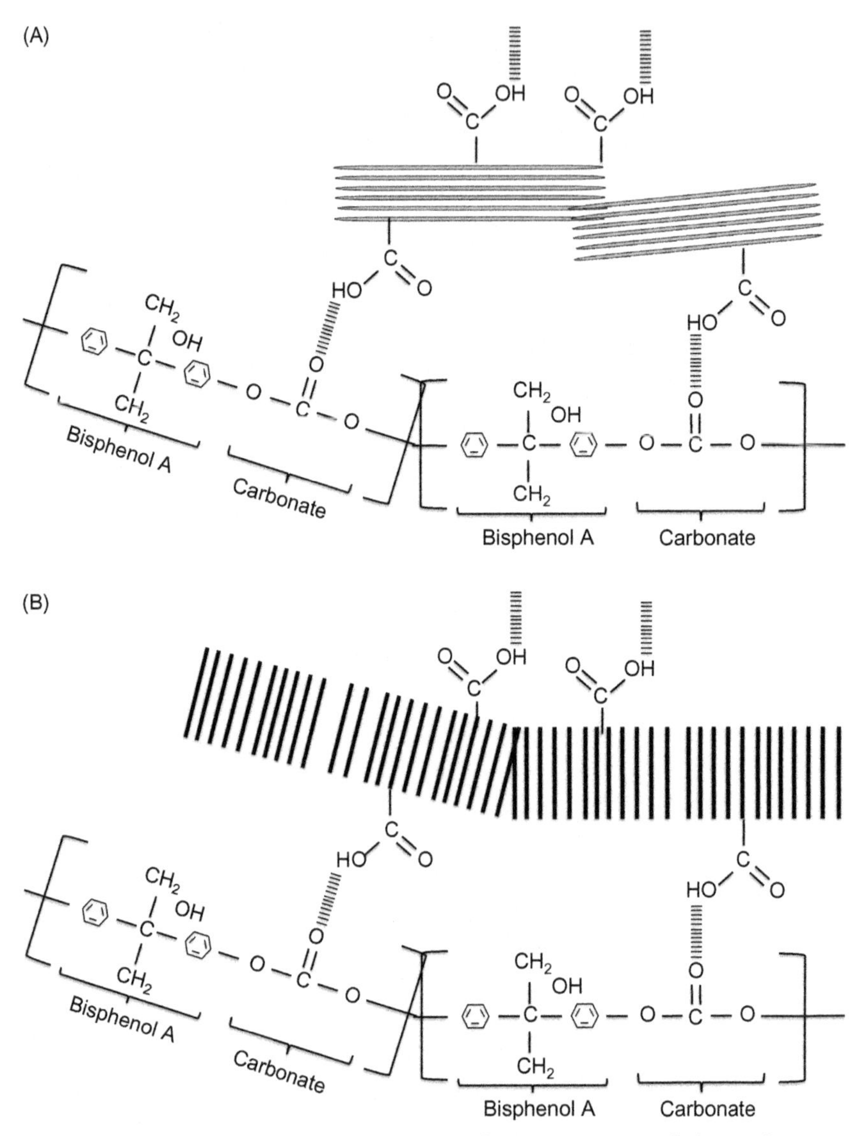

Figure 7.4 Hydrogen bonding. (A) Ribbon GN for improvement of electrical properties. (B) Platelet GN for strength improvement. *GN*, graphitic nanofiber.

Tables 7.1–7.4 list the results of the three investigations:

Economics: The mechanical improvements would probably not be as attractive as the electrical properties enhancements for polycarbonate with all its existing superiority over other polymers in physical characteristics. So improving the electrical properties with GN is worth a look. Like any

Table 7.1 Graphite/PC composite

Solid properties of PC/graphite composites

Concentration		Surface resistance			Gas permeability		Modulus		CTE
					He	N_2	Tensile	Bending	
wt%	vol%	*F* (Ω)	D^2 (Ω)	*B* (Ω)[a]	*F* (Barrer)		*F* (GPa)	*B* (GPa)	*F* (× 105/°C)
0	0	1.3×10^{13}	8.8×10^{13}	5.5×10^{13}	12.5	0.36	2.08	2.58	7.89
1	0.5	2.8×10^{13}	5.1×10^{13}	4.3×10^{13}	—	—	2.16	2.76	7.56
3	1.6	—	7.7×10^{13}	—	10.6	0.32	2.45	—	6.70
5	2.7	—	3.7×10^{13}	3.4×10^{13}	9.32	0.31	2.50	3.21	6.28
6	3.3	—	1.4×10^{13}	—	—	—	—	—	—
7	3.8	3.5×10^{13}	1.7×10^{9}	—	8.61	0.26	2.96	—	5.43
8.5	4.7	7.8×10^{12}	1.9×10^{7}	—	8.02	0.26	3.36	—	5.85
10	5.5	7.3×10^{11}	3.3×10^{6}	7.6×10^{7}	7.29	0.21	3.69	4.23	4.57
12	6.7	1.5×10^{10}	3.4×10^{5}	—	6.82	0.23	4.16	—	4.89
15	8.5	1.6×10^{7}	7.8×10^{4}	—	—	—	5.17	—	3.53

[a]Annealed for 48 h.

Source: From H. Kim, C.W. Macosko, Processing-property relationships of polycarbonate/grapheme composites, Polymer 50 (2009) 3797–3809.

Table 7.2 Thermally treated graphite oxide/PC composite

Solid properties of PC/FGS composites							
Concentration		**Surface resistance**		**Gas permeability**		**Tensile modulus**	**CTE**
				He	**N_2**		
wt%	**vol%**	***F*** **(Ω)**	***D^2*** **(Ω)[a]**	***F*** **(Barrer)**		***F*** **(GPa)**	***F*** **($\times 10^5$/°C)**
0	0	1.3×10^{13}	8.8×10^{13}	12.5	0.36	2.08	7.89
0.25	0.13	—	1.8×10^{13}	—	—	—	—
0.50	0.26	7.2×10^{12}	2.4×10^{13}	11.5	0.33	2.15	7.17
0.75	0.40	—	1.9×10^{13}	—	—	—	—
1.00	0.53	9.8×10^{12}	2.4×10^{13}	—	—	2.22	7.07
1.25	0.66	—	3.3×10^{13}	—	—	—	—
1.50	0.80	7.3×10^{12}	1.6×10^{6}	10.2	0.30	2.27	7.11
2.00	1.06	1.1×10^{8}	8.0×10^{5}	9.7	—	2.51	6.64
2.50	1.33	6.4×10^{6}	6.6×10^{5}	8.9	0.24	2.51	6.73
3.00	1.60	2.8×10^{6}	2.0×10^{5}	8.8	0.20	—	—

[a]Annealed for 48 h.

Table 7.3 Electrical conductivity improvement for oxidized cup-shaped GN

PC/ox-CNF (wt%)	0.0	0.5	1.0	3.0	5.0	7.0
Resistivity (Ω m)	4.2×10^{13}	4.5×10^{12}	9.2×10^{10}	39.3	23.3	12.3
Permittivity at 100 Hz	3.4	3.9	4.3	4.5	7.4	8.7
Time constant (τ)	1.4×10^{14}	1.7×10^{13}	3.9×10^{11}	176.6	172.7	107.01

Source: From S. Kumar, B. Lively, L.L. Sun, B. Li, W.H. Zhong, Highly dispersed and electrically conductive polycarbonate/oxidized carbon nanofiber composites for electrostatic dissipation applications, Carbon 48 (2010) 3846–3857.

Table 7.4 Mechanical & thermal properties of oxidized GN/PC composites

	Storage modulus (MPa) at 140°C	**Young's modulus (MPa)**	**Hardness (MPa)**
0% GN	1700	62	98
5% GN	2200	72	105
10% GN	2700	85	111

Source: From Y.K. Choi, K.I. Sugimoto, S.M. Song, M. Endo, Mechanical and thermal properties of vapor-grown carbon nanofiber and polycarbonate composite sheets, Mater. Lett. 59 (2005) 3514–3520.

of the commoditized plastics, the market for polycarbonate is extremely competitive. The estimated price for general-purpose polycarbonate to a small to medium size manufacturer would be ~\$5.50–\$6.00 per Kg is ~\$2.5/Kg. A 3–5 wt% addition of exfoliated or oxidized GN with a reasonable cost would be acceptable to industry, since the largest users are high value applications such as electronics and automobiles. Nevertheless, the nanomaterials would still have to be much lower in cost to break into this application. Automobile manufacturers would prefer plastics with good ESD qualities for ease of painting. Let us assume the industry is willing to pay 25% more for the enhancements, and let us ignore the actual manufacturing cost. This would put an upper limit of \$1.13 per Kg, ignoring the costs they would incur for blending the GN into the base polymer. At 3% by weight, the increase in cost per Kg would be \$1.50, if we ignore the costs the manufacturer would incur for the actual blending of the GN into the base polymer cost. This represents a 25%–27% increase in cost of materials. At 5%, we will increase costs by \$2.5 per Kg or 41–45%. The justification for substitution would be pretty difficult, but not totally out of the question given the costs an electrically conductive PC would eliminate for the end user, such as the automobile manufacturers. This translates into a a materials cost increase of 60%. At 5% addition, the cost of GN is identical to the cost of polycarbonate resin. The cost benefit of enhancing the electrical properties and replacing another material would have to be amazingly high. We can see the issue with polymer enhancement with nanomaterials now.

There are publications describing the addition of reduced Graphene Oxide (GO) exfoliated by thermal expansion for enhancement of polycarbonate. Given the cost for GO, we will treat these publications only as good to know research for now [272–276].

7.1.3.2 Polyamide composites

Polyamides are polymers with repeating amide units (C=O and N–H). Proteins are naturally occurring polyamides. The most commonly known polyamides are Nylons (Dupont trade name). Nylons are aliphatic amides, whereas aromatic amides are also known (Kevlar). GN with –COOH and –OH groups can form hydrogen bonds with polyamides (Fig. 7.5).

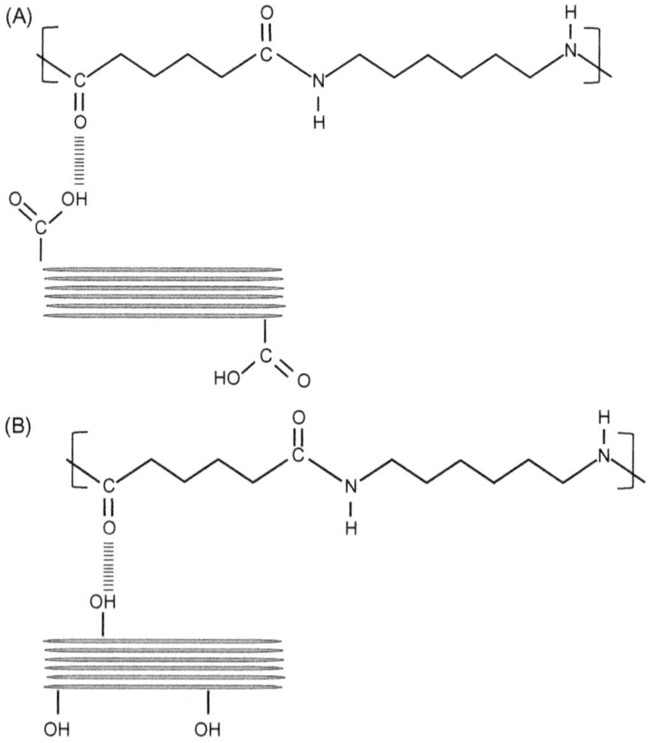

Figure 7.5 Hydrogen bonding with polyamides. (A) Carboxylic with Nylon 6,6 and (B) Hydroxyl with Nylon 6,6.

From a commercial perspective, and my personal investigations, Nylon® 6,6 can be a very successful ESD material for floor coverings. Electrostatic discharges are a nuisance to people and a danger to electronic manufacturing. The commercial world currently adds carbon as fibers as a proportion of the dimension of the carpet fibers to increase conductivity for ESD. This means you can get ESD carpet in any color, as long as it is grey. Lighter color carpets may be possible with low percolation thresholds. Leer et al. [277] determined a percolation threshold of 10%. The high percolation threshold is indicative of an earlier discussed consequence of agglomeration and clumping. Unless some new approaches are investigated, this would be a no go.

Currently available research and investigations do not indicate that strength and other mechanical properties are improved significantly enough to justify mention [277,278]

7.1.3.3 HDPE composites

Improvements in performance of HDPE composites [279] and polypropylene composites have been investigated by physical dosing means [280–283], but here again, the marginal improvements reported don't qualify this process as a good route for commercialization.

7.1.3.4 PLA & PGA

Polylactic acid and polyglycolic acid type of aliphatic (main chain) polyesters are becoming increasingly popular as biodegradable and bio-based plastics. The melting temperature of PLA can be increased by 40–50°C and its heat deflection temperature can be increased from approximately 60°C to up to 190°C by physically blending the polymer with poly-D-lactide. PLA is currently used in a number of biomedical applications such as drug delivery devices. This high value market holds great promise, and PLA can gain wider acceptance if strength and conductivity could be imparted with low dosages.

The value addition to PLA would be substantial with GN enhancement (see Fig. 7.6). The applications are in high demand due to the biodegradability and food grade status. The plastic could be more useful with improved physical characteristics. The current price is around $1.0/lb. However PLA is used in high-value markets, so a higher price may be palatable. For example, "Graphene" enhanced PLA is commercially being launched by Haydale Graphene Industries (UK). The target market is three-dimensional (3-D) printing. 3-D printing market is

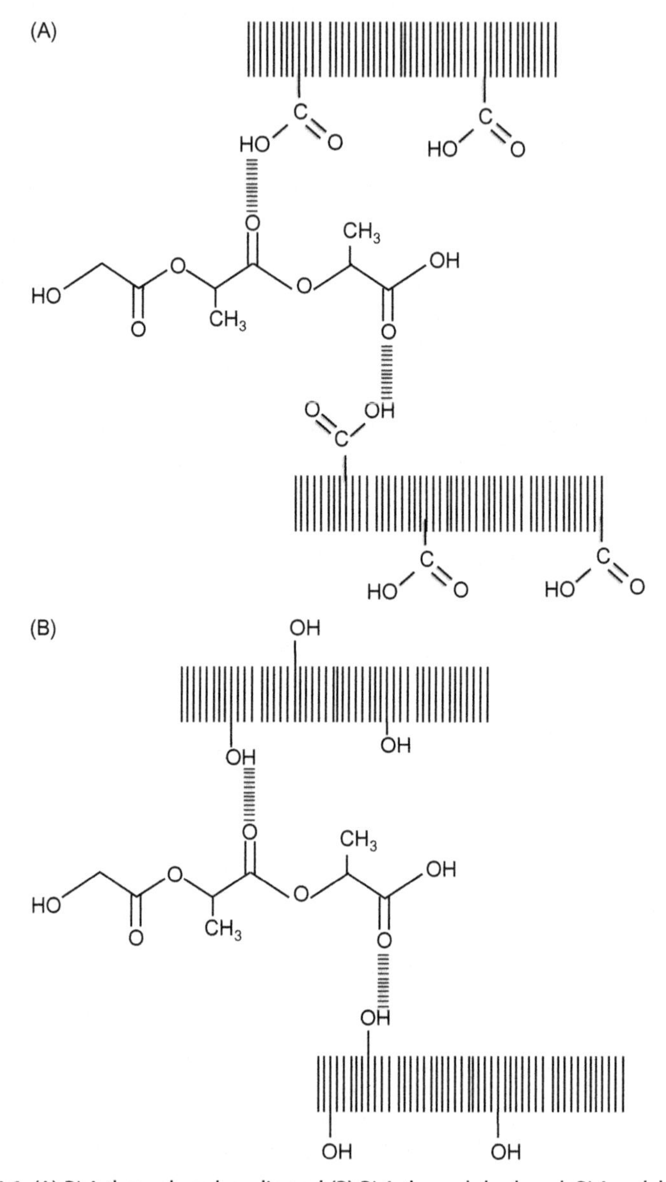

Figure 7.6 (A) PLA through carboxylic and (B) PLA through hydroxyl. *PLA*, polylactic acid.

going through explosive growth, and enhanced materials certainly will be desired to increase market share. Black Magic 3-D sells it on their website for $550/kg. That is a decent jump from $2.2/kg for virgin PLA. Let us hope the product starts selling in high volumes. The market is certainly big enough.

7.1.3.5 Coefficient of thermal expansion (CTE)

Graphite has long been known to have low thermal expansion [284], and the coefficient of thermal expansion of GN-epoxy composites has been shown to be substantially lower than that of the matrix [285]. From the cited work with GN, we may presume that CTE reduction will be another field where GN may contribute to desirable improvements.

When a materials specialist considers replacing a metal part with a composite, thermal and electrical performance may often be of critical importance. Inadequate heat dissipation, EMI shielding, and electrical grounding will often prevent exchange of a lighter composite for a heavier metal part, even though weight-sensitive niche applications can often tolerate the higher nano composite cost. The aerospace industry in particular has many such applications—particularly in satellites. The competing technology, the addition of metal powders, may result in unacceptable corrosion and impact strength degradation. Prototypes incorporating GN have been fabricated from thermoplastics, thermosets, and elastomers. Although it would be desirable to achieve fully multifunctional performance, yielding improvements in mechanical properties as well as thermal and electrical characteristics, there are many applications where the thermal and electrical benefits alone justify the use of GN. The necessity of going beyond conventional compounding techniques has created opportunity for the emergence of a supply chain producing VGCNF (CVD-GN) compounded materials. Resins are now available that contain 20% by weight loadings of GN that can be utilized in a variety of formulated epoxy systems including pre-pregs, molding compounds, adhesives, and coatings [286]. Such formulations are intended for use in conductive adhesives with high strength characteristics, structural composite panels to replace metal for weight savings and corrosion resistance, and components for medical, aerospace, and electronics applications. Use of GN has been reported for improved mechanical properties of liner-less composite pressure vessels, where performance improvement is attributed to the development of high strain, microcrack-resistant resins and the inclusion of GN at the ply interface. These materials eliminate microcracking as the first failure mode and improve the laminate failure strain to a level that nearly equals that of the Toray T700 reinforcing fiber [287]. Finally, the availability of a thermal grease composed of a silicon-free carrier material incorporating GN as the thermal conductor has also been announced [288]. Here the nano-scale filler enables accurate thin film application and decreases thermal resistance.

7.2 LITHIUM-ION BATTERIES

7.2.1 Background

Li-ion batteries do not need any introduction. They are all around us in consumer electronics as a replacement for the long reigning rechargeable Ni–Cd batteries. Commercial lithium-ion batteries with high-energy density and good cycle ability usually consist of a graphite-based negative electrode, a positive lithium metal oxide electrode, and a separator soaked with an organic electrolyte. The roles of the electrodes reverse between anode and cathode, depending on the cycle (charge/discharge). High quality graphitic material is preferred as the negative electrode due to the gap between the basal planes. Li ions intercalate within these planes back and forth, depending on the cycle. The positive electrode can be lithium iron-phosphate, lithium magnesium oxide, or lithium oxide.

Despite the rapid pace of development of Li-ion batteries, several issues remain unresolved satisfactorily. One of them is the rather high irreversible charge loss occurring after the first charging of the graphite electrodes. It is generally accepted that this charge loss is mainly due to reductive decomposition of the electrolyte on the negative electrode [289]. The resulting protective film, called solid electrolyte interface (SEI), allows lithium-ion transfer but impedes electron transfer [290,291]. The SEI formation is also proposed as the main reason for Li-ion battery's poor performance in cold temperatures [292]. It has been reported that at (−40°C) a commercial Li-ion battery only delivered 5% of energy density and 1.25% of power density, as compared to the values at (+20°C) [293]. The performance issues associated with Li-ion batteries can thus be summarized as

1. Reduced ionic conductivity of the electrolyte and SEI formed on the graphite surface [294–296].
2. Limited diffusivity of lithium ion within graphite [297,298]
3. High polarization of the graphite anode, which is associated with the former two factors [299].

From the above, we can infer that the hurdles associated with Li-ion batteries are materials related. Specifically, the materials used at the anode. Flake or amorphous graphite of high crystallinity is currently employed for this application, such as Highly Oriented Pyrolytic Graphite (HOPG). Fundamental to the issues is the irregular structure of graphite and the high-energy dangling edge sites with armchair and zigzag faces.

Zhao et al. [297] and Peled et al. [300,301] have done in-depth studies of the formation, location and intercalation effects of SEI. Based on thermo gravimetry-differential thermal analysis combined with mass spectrometry (TG-DTA/Ms) and XPS, they report some very enlightening observations. Details of the composition of SEI on the edge of the graphite (cross section) and on the basal planes were shared. The lateral distribution of SEI forming compounds at sub-micron resolution found that mainly electrolyte breakdown products like Li and F and some polymers like polyolefins and oxygen bound polymers were deposited on the armchair and zigzag dangling edges. The basal SEI deposit was found to be mostly polymeric materials (50%) and $LiCO_3$ (10–30%) indicating solvent reduction to be the greatest contributor for the deposit. Alkoxides and Li_2O were also found in the bulk, mostly at the bottom of the basal SEI deposit. Deposit thickness was determined by the carbon signal. In the basal region, the SEI thickness was 2 nM, whereas at the edges the thickness was more than 30 nM, creating a much larger hindrance to Li-ion intercalation.

SEI plays a major role in the performance of the battery and acts differently on the basal plane than it does on the edges. On the basal plane, it is polymeric, so forms a very shallow nonconducting film, but on the edges, it must be very conductive to Li ion. The discussed observations now shed some light on the three factors listed above:

1. Reduced ionic conductivity of the electrolyte and SEI formed on the graphite surface. A 2-nM thick layer on the basal plane and a 30-nM layer on the cross section would contribute to this state. We can deduce from the earlier discussion that on the basal plane, the depth of Li-ion intercalation may be affected and charge/discharge rate may be affected if the basal plane is wide and requires irregular and long distances.
2. Limited diffusivity of lithium ion within graphite. Now shown to be a function of the thickness of the SEI on the edge (cross section) of the graphite, if the SEI on the edges is thick, and contains nonconductive organics, the diffusivity will naturally suffer.
3. High polarization of the graphite anode, which is associated with the former two factors. Now it can be seen that the polymer and ionic SEI layers will create the polarization effect, slowing down charge/discharge cycles.

GN materials could possibly be uniquely qualified for use in lithium-ion anodes (Fig. 7.7).

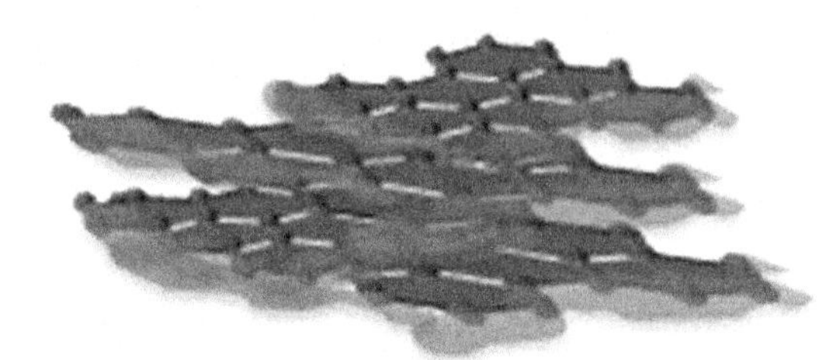

Figure 7.7 Disordered graphite edges with SEI deposits. *SEI*, solid electrolyte interface.

1. GN have very a high specific surface area. Platelet type and ribbon-type GN can be employed for Li-ion applications. The edge faces are exposed with very low depth (width of the GN). As we already know, the edges are extremely active and possess high energy levels, but can be brought down to their lowest energy level by heat treatment (1800°C), which seals the edges by forming loops. The new low-energy state does not allow formation of a very thick layer of SEI, and high diffusivity is maintained, see Fig. 7.8A and B.
2. GN have very small basal planes, but these planes are of consistent size, making it physically impossible for the SEI deposit to be formed over large areas, consequently eliminating the physical hurdle for Li-ion to and from the active regions.

Mariguchi et al. [298] have reported work done with molecular dynamic (MD) simulations, which confirm my explanation of the higher performance of the GN with closed loops, when applied to Li-ion batteries. They call these graphite polyhedral crystals (GPC). They show very low capacity loss in Li-ion battery anodes, when heat-treated GN were used as anodes. They confirm with MD simulations that the thermal vibrations of the free edges play a key role in the formation of the loops. Munetoh et al. [299] detail a more comprehensive investigation into the same phenomenon. Using a cluster model, with energy optimization defining the clusters, they have showed that Li^+ diffusion occurs more readily in the Y direction (perpendicular to the C—C bonds) rather than in the X direction along the C—C bonds. The diffusion barrier is calculated for a Li ion between two graphene planes to determine the least energy diffusion path for Li ions to the GN anode. The calculated value for Li diffusion barrier height is about 1.6 eV. They present a remarkable argument that the Li diffusion path to the anode is from the "open interstices" between the closed loops of the GN as follows: Li^+ reaching the

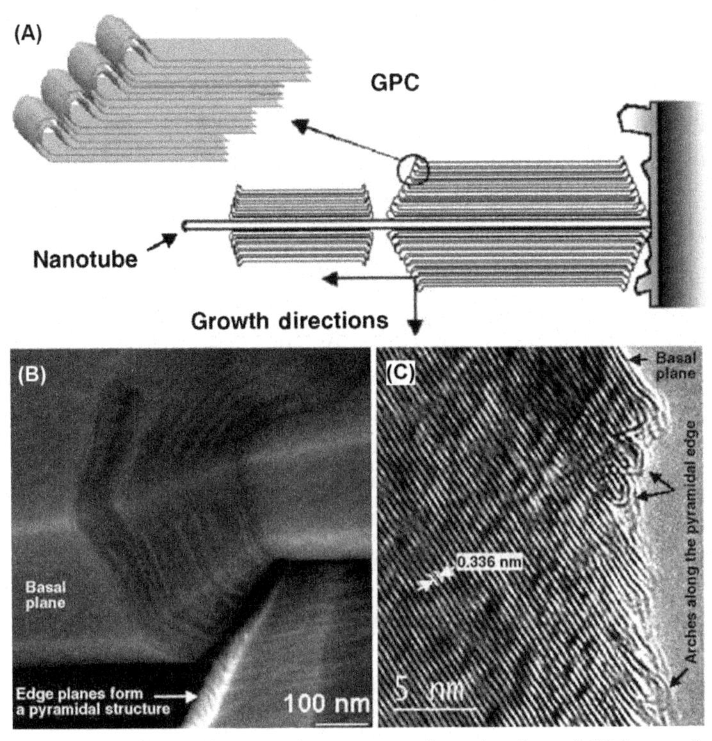

Figure 7.8 (A) Schematic showing the suggested mechanism of GPC growth around a carbon nanotube core; (B) SEM micrograph of pyramidal surfaces of two GPCs and (C) TEM micrographs of a GPC tip showing arches formed by folding graphene sheets [22].

outermost wall of the closed loops diffuses to the entrance of the open interstice along the curved graphite surface. The diffusion barrier in this case was estimated to be about 0.6 eV with their model approximation. Because the internal region from the entrance of the open interstice is constructed of the layered structure, the Li diffusion barrier was estimated to be about 1.6 eV. Therefore, they concluded that the Li^+ intrusion barrier to graphite anodes through the open interstice roughly corresponds to this value of 1.6 eV. See Fig. 7.8C.

Next, they show the results of Li^+ intrusion barrier through ring defects in the closed loops. Li^+ intrusion barrier heights through six-membered ring, seven-membered ring, and the symmetric vacancy exceed more than 5 eV. Therefore, it would be impossible for Li^+ to pass through these ring structures. In contrast, the barrier height for the vacancy-type ring is calculated to be only about 1.0 eV, which is compared with the corresponding value of 1.6 eV for the open interstice as shown previously. Through these calculations, they conclude that the Li-ion intrusion sites on the surface of the GN anodes with closed loops are assumed to be the open interstice and/or the void defects. Interestingly, they note that if the number of intrusion sites increases, the lithium ions are predicted to be stored smoothly in the anode material. This desired quality is well provided by closed looped GN, which will contain very high numbers of small, graphene planes with 3−4 layer closed loops, providing large numbers of interstice openings.

Finally, Mariguchi et al. [298] show that the closed loops, because they prevent SEI formation, are also responsible for remarkable charge/discharge properties, having very low irreversible charge characteristics. See Fig. 7.8C.

Indeed, Yoon et al. [299] have shown the GN to perform at higher efficiencies than graphite and CNTs over more than 10 cycles. They also reported data showing low temperature (−45°C) performance of >97% with Polycarbonate (PC) electrolyte, versus other materials not showing any performance at all.

Anode material	**Electrolyte**	**Capacity (mA/g)**		**Efficiency**	
		70	350	3rd	10th
GN	EC + PC	355.7	342.1	93.8%	98.1%
	PC	339.1	320.3	93.0%	97.5%
Graphite	EC + PC	316.9	298.7	92.6%	98.3%
		0.00	0.00	0.00	0.00

Source: Baker et al. Catalytic Materials, LLC.

The advantages given above would increase energy density of Li-ion batteries. The current prices for Li-ion battery grade graphite is in the range of GN. GN should be seen as high quality graphite in this case, because it is highly crystalline and organized. Pushing "nano" in the nomenclature simply puts up a wall, even with the early adopters, because the hype has not delivered anything so far. If explained as a highly crystalline, ordered graphite, perhaps there may be a change in attitude, and a partnership could be formed between an early adopter and a manufacturer.

7.2.2 Summary

Obstruction of the Li-ion intercalation within the anode is theorized to be the main reason for the loss of recharging capacity of Li-ion batteries after the first charge. The occlusion is shown to be due to a polymer buildup at the graphite edge because of a reaction between the edges and the electrolyte. GN with the edges sealed by heat-treating, provide a nonreactive edge, with a unique diffusion barrier sufficient for the Li-ion to intercalate through the space between the closed loops. A decrease in weight for the same charge capacity increases the energy density. Approximately 30% of the weight of the battery is the anode, so a significant benefit could be achieved by reducing the weight of the anode. The findings certainly deserve more investigation. As for the prices, Li-ion batteries can afford the price, if the numbers from Baker et al. are accurate. The increase in performance, especially cold temperature performance, would be well worth the relatively small increase in cost.

7.3 CATALYSIS

Catalysis is the primary reason I got involved with nanomaterials more than 15 years ago. A friend proposed activation of conventional catalysts electrically. The original thought turned out to be simply a localized heating technique, but it triggered an ardent desire in me to study the implications of electronic perturbations on catalytic activity. With their unique qualities, nanomaterials provided a means to an end for my quest to try a quantum mechanics approach to catalysis. 15 years ago, with not an iota of academic background in catalysis or nanomaterials, I naively jumped head first into this vast and complicated field with such a simple and crazy idea. However, I had the advantage of lacking tunnel vision, and was able to achieve some of my goals. My limited experience and knowledge gained from perusal of hundreds of publications in this field tells me that for the medium term, catalysis is probably the best prospective field of application for GN. No doubt, the materials and their performance benefits would have to go through the lengthy process of evaluation, validation, and product introduction cycles, but given the proof of performance and low manufacturing costs, the process can start in the short term (2–3 years), reaching commercialization in the middle term (5–6 years). Undoubtedly, in catalysis, when you get to the decision node about the possibility of utilizing a better performing material already available in the market, you would conclude in the negative. If assured a sustainable supply, you would go forward with the new path of incorporating nanotechnology in your process. The GN cost and technology is at a point that the research work presented below can be moved forward to commercial pilot scale. The improvement in performance, if proven on an industrial scale, would have a direct impact on "profits" in the commercial world, leading to greener processes as a side benefit. As a first step in the application of GN in the field of catalysis, I suggest viewing these materials simply as macroporous supports, facilitating a simple "substitution" approach to replacement of current catalyst supports used by industry. As I mentioned in the Li-ion section, one of the side effects of the hype about new materials is the notion it injects in people's minds that the materials are fundamentally different in chemical composition from other materials being used today. GN are simply a more uniform and higher purity version of graphite. There is very little apprehension in trying out materials that honestly claim that they are, say a better version of graphite, such as HOPG. First problem is, almost everything done with carbonaceous nanomaterials today is somehow commercially defined with the term graphene. Then, graphene has been put on a pedestal so high that it triggers a perception of an inherent risk involved in taking the first step. The focus should have been cost reduction of what we already have, such as CNTs and GN. Remember how expensive Teflon was, or composites were when first announced? In this section, I examine some of the applications that I believe are ready for the move forward, as well as some that have potential, but require more development, by analyzing some publications that claim otherwise.

A relatively important point to note for the applications shared below is that some of the GN discussed in the applications by the researchers below were synthesized from hydrocarbon gases other than methane and syngas. Probably the reason was the ease of synthesis given previous work done by others, since the focus was on the applications. GN manufactured using methane, as a feedstock with the appropriate catalysts would perform just as well.

Designing and developing a highly selective catalyst has been an intriguing and challenging goal in catalysis research. The development of novel industrial catalysts ideally begins with the identification of a catalytically active species and an appropriate catalytic supports. Over the years we have learnt that elemental carbon exhibits properties of both an active catalytic species and of a catalytic support. The catalytic properties of carbon, however, are less known compared to traditional metal–oxide catalysts. I will go through a few applications below that use carbons as catalysts or substrates, with the intent of comparing their performance with that of GN. Activated carbon is a commonly used substrate in catalysis. The high surface area afforded by the micropores, mesopores or macropores generated by activation is ostensibly an attractive feature for better dispersion of the catalyst metal. Micropore volume is highest in activated carbons with very high ASTM (American Society for Testing and Materials) hardness, such as coconut shell carbons. The hard material allows for very thin pore walls that don't collapse easily. Accordingly, mesopores are generated from materials such as coal and macropores by soft materials such as wood, peat, etc. The type of activated carbon used for a particular catalytic process depends on the attrition expected, diffusion limitations due to size of reactant molecules and catalyst metal percentage loading. Recycling catalyst metals, especially precious metals is easier because activated carbon supports can be vaporized by oxidation, along with some of the promoters, leaving relatively pure metal. Unfortunately a very small percentage of these pores are truly accessible by the reactant gases/liquids in actual operations. Diffusion barriers may lead to a "retardation" effect on one of the reactants, resulting in less than optimal performance. It is well known that in liquid phase reactions, the diffusion of the reactants through the solid catalyst matrix and back diffusion of the products are significantly influenced by the external size of the particles [302].

In catalysis, GN act as conductive substrates that can cause significant electronic perturbations [303]. The catalysts are deposited on either the armchair or zigzag edges of the GN, creating unique interactions between GN and the metal. In addition, their physical characteristics such as a lower diffusion barrier due to external surface area and conductivity give them many advantages over the current substrates in most of the catalytic reactions.

As-produced GN with reactive edges can often be used directly as catalyst supports. The conductive GN supports present clear differences with respect to activated carbon, and a recent theoretical study related to the interaction of transition metal atoms with GN indicates major differences [304] (Fig. 7.9).

For some applications, the binding sites are dependent on the structure of the support: the studies conducted over nickel show that the curvature of CNTs could provide more stable anchoring sites. In such instances, we would use heat treated GN, because of the curvature of the loops formed at the edges when GN are heat-treated (discussed previously). The GN would also induce strong modifications of the electronic properties. A comparison of SWNTs to crystalline graphite shows a modification of the π-electron cloud [305,306]. The curvature also affects the values of magnetic moments on the nickel atoms on

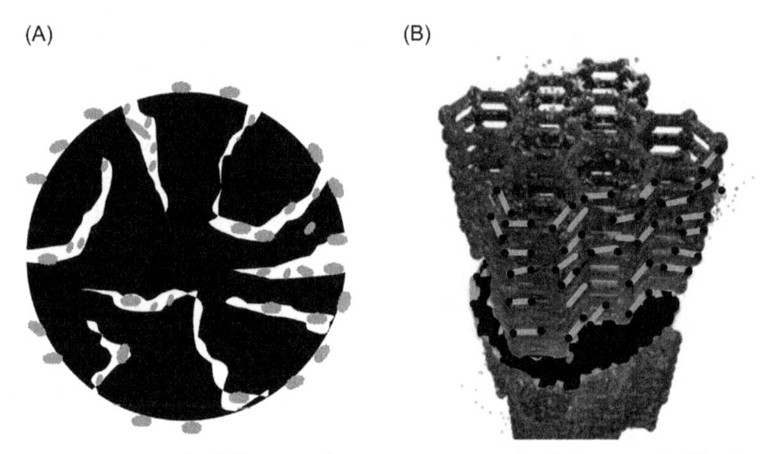

Figure 7.9 Structural difference between Activated Carbon and Platelet GN. (A) Activated carbon—(typically these are not perfect spheres). Catalyst particles deposit on surface and within pores as globules. (B) Platelet GN—catalyst particles deposit on edges, as flat surfaces. *GN*, graphitic nanofiber.

the tubular surface walls and the charge transfer direction between nickel and carbon can be inverted [303]. Therefore, a strong metal substrate interaction (SMSI) also exists in this case. SMSI induces a spreading of the metal on the support surface. In contrast, globular particle geometry is prevalent when nickel is supported on γ-alumina, which is consistent with the existence of a somewhat weaker metal–support interaction.

GN have also been directly grown on carbon felts and fabrics and other supports [307]. GN as wash-coats on ceramic supports have also been reported [308].

In some cases discussed below, where SWCNTs are deemed to be good supports, heat-treated GN have been suggested as a lower cost alternative with comparable performance with the additional advantage of being able to produce them with consistent physical structures and properties. (SWCNTs cannot be made with consistency at the current time, and as such, are confined to the research world.)

7.3.1 Oxidation of ethylene to ethylene oxide with GN-supported Ag catalyst

In industry at present, Ethylene is partially oxidized with oxygen over a silver alumina catalyst to form ethylene oxide (EO) [391].

$$CH_2 = CH_2 + \frac{1}{2}O_2 \rightarrow CH_2 - CH_2O \quad (\Delta H = -106.7\ \text{kJ/mol}). \quad (7.1)$$

Ethylene can be completely oxidized to form carbon dioxide and water.

$$CH_2 = CH_2 + 3O_2 \rightarrow 2CO_2 + 2H_2O \quad (\Delta H = -1323\ kJ/mol) \quad (7.2)$$

The product Ethylene Oxide can also get further oxidized to form carbon dioxide and water.

$$CH_2 = CH_2O + \frac{5}{2}O_2 \rightarrow 2CO_2 + 2H_2O \quad (\Delta H = -1323\ kJ/mol). \quad (7.3)$$

Background: EO is an important chemical, which is used as an intermediate in the production of glycols and plastics. EO is obtained industrially through partial oxidation of ethylene on silver catalysts. EO is one of the most important raw materials used in large-scale chemical production. A total of 75% of the world production of EO is used for synthesis of ethylene glycol (EG) including di-EG and tri-EG. It is used as a disinfectant, in polymerization reactions and many other applications.

Industry uses either air or oxygen as feed streams to obtain an oxygen source to react ethylene with oxygen. The oxygen-based process is chosen due to its many advantages. For all plant capacities and a given type of catalyst, the oxygen-based reactor yields a higher selectivity and requires less catalyst. Although the air-based process may cost lower to run, the initial building costs of the air-based plant is much more than the oxygen-based plant. While the oxygen-based process requires a carbon dioxide removal section, more stainless steel, and some expensive instrumentation, the air-based process requires more catalyst (50%), more reactors, a multistage compressor, air purification units, and a vent gas treating system.

If a small improvement in yield can be established by a new catalyst substrate, it would be an attractive proposition for the manufacturers of this commodity chemical.

The oxidation of ethylene can lead either to the desired product (EO) or to the thermodynamically favored acetaldehyde (Ac), which is then readily converted into carbon dioxide. The optimal catalysts for this reaction should therefore selectively promote the formation of EO while avoiding total combustion.

The only catalyst for ethylene epoxidation used in the industry is silver, typically supported on alumina. Pure silver gives selectivity values toward EO of around 40%, but the use of suitable promoters can enhance the selectivity to near 90%. The reaction mechanism is still not known, and in particular which oxygen species is responsible for the selectivity is widely debated. The silver-oxygen system is the focus of much investigation. A variety of dynamically evolving oxygen species have been shown to coexist on silver surfaces [310].

Electronic structure calculations on the chemisorption of atomic oxygen on Ag(1 1 0) and on the subsequent reaction of this chemisorbed oxygen with ethylene show that the presence of subsurface oxygen (1) reduces the bond energy between silver and adsorbed oxygen and (2) converts the repulsive interaction between adsorbed oxygen and (gas-phase) ethylene into an attractive one, thus facilitating the reaction. The presence of subsurface oxygen diminishes an important four-electron destabilizing interaction (Pauli repulsion) [311].

Xu et al. [309] studied the performance enhancement of the reaction by use of platelet type GN, heat-treated GN (treatment in Argon at 2300°C), pristine crystalline graphite, and α-Al_2O_3. The GN were synthesized by using CO and H_2 as the feed gas. The experiments were performed at atmospheric pressure (commercial production uses >10 atm.). The modified (heat treated) GN displayed a significant increase in performance, 150% higher activity (2.5 times), much higher selectivity and yield with the GN supported Ag. In addition, the selectivity of the GN supported catalyst improved with reaction time, whereas the conventional catalyst support declined in activity over time. They report the following results at atmospheric pressure and claim the performance improvement would be similar at the commercially used pressures (Table 7.5).

The influence of subsurface oxygen on this reaction may be one of the factors being enhanced with modified GN as a substrate. The

Table 7.5 Results reported by Xu et al. [309]

Catalyst	O_2:C_2H_4:Inert	% Selectivity	% Conversion	% Yield
Ag/GN	3:12:85	49.90	8.0	3.98
Ag/modified GN	3:12:85	39.57	25.48	10.08
Ag/α-Al_2O_3	3:12:85	30.60	10.30	2.67
Ag/pristine graphite	3:12:85	22.94	4.52	1.04
Typical commercial Production parameters and results With shell direct oxidation process At 250°C and 200–300 psig (industry average)	1:1	80–85	8–10%	6.4–8%

electronic structure of Ag on heat-treated GN may have some electrochemical relationships worth exploring. The improvements obtained certainly warrant an investigation of the modified GN at higher (commercially used) pressures. It may be possible to achieve commercial level performance at slightly lower pressures. This change in operating conditions would save the manufacturers energy costs, translating into a more competitive position in the market place.

As always, cost considerations would be important. The cost of producing GN is accompanied here by the heat treatment of the GN at high temperatures. The required temperatures can only be achieved by (expensive) electrical heating. The cost of inert gases such as argon would also play a role in the calculation, because during this treatment, some residual materials used in the catalyst may vaporize, and the inert gases would have to be replaced. The furnace capital costs will be higher, and the temperature in batch processes would have to be lowered for removal of the material, or alternatively, the GN must be cooled before unloading, necessitating energy considerations.

At first glance, and compared to the other substrates, the high temperature heat-treated GN performed much better. This reaction therefore seems to point to and partially validates the hypothesis according to which, the curvature of the graphene backdrop of the catalytic site is important to the electronic perturbation discussed previously. The EO selectivity for the reaction would have to be much higher at elevated pressures to match the economics of the incumbent processes. In industry, the unreacted ethylene is typically recycled back to the process, whereas the reacted ethylene, not ending up as product, is discarded or used a fuel replacement. The cost of ethylene is the most expensive part of the process. Therefore, low conversion with high selectivity is preferred over high conversion with low selectivity. Let us look at the results from a business perspective:

In this case:

$$\text{Conversion} = \frac{\mathbf{n}_{C_2H_4\ \text{in}} - \mathbf{n}_{C_2H_4\ \text{out}}}{\mathbf{n}_{C_2H_4\ \text{in}}}$$

Selectivity is defined as

$$\text{Selectivity} = \frac{\mathbf{n}_{C_2H_4\ \text{out}}}{\mathbf{n}_{C_2H_4\text{in}} - \mathbf{n}_{C_2H_4\ \text{out}}}$$

$$\text{Yield} = \frac{\mathbf{n}_{C_2H_4\ \text{out}}}{\mathbf{n}_{C_2H_4\ \text{in}}}$$

Looking at yield in another way: yield = conversion × selectivity

Assuming unreacted ethylene is recycled, and the only other by product is CO_2, we may come to the following conclusion for the modified GN substrate:

$$25.48\%(\text{conversion}) \times 40\%(\text{selectivity}) = 10\%\ \text{EO yield}.$$

But 60% of the converted ethylene ended up as CO_2 ($25.48\% \times 60\% \sim 15\%$) so 15% of the feed ethylene is lost as waste, mainly due to low selectivity.

Conceptually, all the unreacted ethylene would be recycled. So if the conversion was lower, more of the feed would be recycled. In the above calculation, we see that 10% of the feed yields the product EO. 15% goes to waste, so 75% of the feed stream is now recycled. For an economic evaluation, we would consider the 15% loss as part of our cost equation.

Comparatively, by using the same procedure, in the incumbent (industrial) process we see 10% (conversion) $\times$ 80% selectivity = 8% EO yield, but only 2% of the feed ethylene is lost as CO_2.

The comparison shows how research data can be confusing for real world applications. Naturally, one would highlight the positives or there would be no point in publishing the paper, but when benefits for industrial use is implied, a complete picture helps. Nevertheless, the results warrant a pilot test at higher pressures. Further investigation would be very beneficial to the industry because there is a salient benefit not mentioned by the authors and one I have personally experienced in the real world with exothermic reactions; all performance results being equal in short run times typical of such benchtop studies, GN would exhibit higher performance over long production runs. The reason being ethylene oxidation is a highly exothermic reaction and under the reaction conditions, GN has 4 times the thermal conductivity than α-Al_2O_3. Heat removal rates are important to avoid catalyst sintering. Notably, then, GN would bring dual advantages for this reaction. First, by nature of the catalyst deposition on GN, there is minimal, if any, sintering. Second, the conditions responsible for sintering, primarily high local temperatures, are eliminated. Silver, the currently used catalyst, is expensive, and even a few more months of life afforded by a new substrate would be bring financial benefits to the manufacturers. To summarize, this publication helps us understand the importance of taking the complete picture into account when implying breakthroughs that claim major benefits to industry in general. On one hand, ostensibly the performance improved, but from a review with real world importance, absolute performance was not viable. On the other hand, another significant aspect not reported, outlines benefits that warrant further investigation. Thermal management in exothermic reactions is easier with thermally conductive substrates. Beyond these factors, my further optimism for the use of GN in this particular application is a consequence of personal experience in using GN in similar reactions where the desired product is technically an intermediate, and the goal is to keep the reaction from achieving complete oxidation to CO_2. The easy accessibility of the reactants to catalyst surfaces when deposited on GN edges, and minimal resistance for the product and unconverted reactants to diffuse away, allowed our team to design very short contact time reactors. Hence this and other similar reactions such as the exothermic catalytic partial oxidation (CPOX) of methane, would be high on my list for pilots if I was still promoting supply of GN as raw materials. It is impossible to do a financial analysis with the data reported in this particular publication, but intuition tells me the differential between $51/Kg cost of the GN and current substrates should be very easy to justify, if even the slightest bit of performance enhancement can be demonstrated.

7.3.2 Fischer–Tropsch synthesis with cobalt catalyst

This is one of the applications I strongly believe can have immediate use for GN, and one I currently am actively involved in. Small-scale FT plants are in their infancy and GN can offer much relief from the major issues that have stymied their progress. *Background:* The Fischer–Tropsch (FT) reaction was first used industrially in Germany during WW II. It converts synthesis gas ($CO + H_2$) obtained from the gasification of coal or biomass, from reforming of methane (biogas, landfill gas or natural gas), to synthetic oils and waxes. The products produced are virtually free of sulfur and nitrogen compounds. The FT process offers an option to monetize gas fields that are not viable. Unlike crude oil, natural gas must be compressed and delivered by pipeline or converted to liquified natural gas (LNG) at the site and transported as a liquid. LNG requires cryogenic processing and is expensive on a small-scale. Around half of all worldwide natural gas resources are remote or stranded in abandoned wells with reserves that are not economically accessible by either pipelines or conversion to LNG. Natural gas is also produced as "associated" gas, when oil is pumped. Consequently, large volumes of natural gas, specifically methane, are flared at oil wells around the world. Methane trapped in coal seams (CSM) or at the interface of coal with water high in bicarbonates (CBM) is also released during coal mining or targeted methane recovery which uses water extraction to reduce hydraulic pressure. Coal associated gas is also dispersed around the world in small volumes which are often uneconomical to transport.

Small capacity FT plants would be required to realize this objective, but such plants have not proven viable yet. 3000 barrels per day capacity and >$60 per barrel oil price are said to be the points at which these plants are viable for liquid fuels. However, other factors such as government credits, subsidies and rising demand for sulfur free fuel may make smaller plants more feasible. Many large plants cannot justify producing fuels because of the fluctuating energy prices. They are mostly focused on value added products such as drilling fluids (DFs), lubricants, and waxes for industrial use.

During the FT process CH*n* monomers, formed via hydrogenation of adsorbed CO on transition metals, produce hydrocarbons and oxygenates with a broad range of chain lengths and functional groups. The major products are linear paraffins and α-olefins.

$$(2n+1)H_2 + nCO \rightarrow C_nH_{2n+2} + nH_2O$$

The products from FT can be tailored to predominantly be within a relatively narrow alkane range. The recent regulation for low sulfur fuel and the global drive to renewable fuels has brought the FT process back in the limelight for fuel production. Small-scale plants have not yet been successful due to various operating difficulties, some of them being catalyst life, thermal management for the exothermic reaction, and energy consumption in the generation of the syngas feed. Economies of scale make the very large plants viable, but the capital investment is in tens of billions of dollars. Large plants are currently operating in South Africa—Sasol (coal gasification to diesel), Shell in Malaysia, and one in Qatar. Sasol and Chevron have a joint venture in Qatar.

The FT synthesis has been recognized as a surface-catalyzed polymerization process via carbide formation on cobalt, ruthenium and iron catalysts [312]. The theoretical conclusions and interpretations of experimental results have evolved since the original mechanism proposed by Fischer [313–324], partly from availability of modern surface science instruments. Davis [325] concludes from experimental data that the initial step in the chain propagation is the formation of an oxygen containing structure.

Therefore, the species responsible for the initiation step are different from the ones involved in the propagation step, which are derived predominantly from CO. They emphasize that the C–O bond is not broken prior to the addition of the CO to extend the growing chain by one more carbon. See Fig. 7.9A.

Further, Storstaer et al. [326] have presented calculations for rate constants using the transition state theory and Arrhenius plots. They present an argument about the elementary pre-exponential steps of the C_1 and C_2 formation. The activation barriers were found to be: Hydrogenation of CO < hydrogen assisted CO dissociation < direct CO dissociation, leading to two sets of elementary formations of the C_1 and C_2 alkene and alkane formation. The "carbide mechanism" would be the hydrogen assisted dissociation of CO and the other based on CO hydrogenation.

FT synthesis can be achieved at high temperature ~300–350°C (HTFT) or low temperatures (LTFT) ~250–280°C. HTFT uses an iron-based catalyst. This process was used extensively by Sasol in their coal to liquid plants (CTL). Higher temperatures favor higher conversion, but the chain formation reaction also terminates faster and methane formation is favored. Higher pressures result in higher conversions as well as faster chain propagation, favoring long chain alkane formation. Typical pressures used are between 20 and 30 atmospheres. The marginal gain in conversion rate with further increase in pressure does not justify the additional cost of capital equipment. Low-Temperature Fischer–Tropsch (LTFT) is operated at lower temperatures and uses an iron or cobalt-based catalyst. This process is best known for being used in the first integrated GTL-plant operated and built by Shell in Bintulu, Malaysia. Most of the metals of the VIII group of the Periodic Table can be used in hydrocarbon synthesis from carbon monoxide and hydrogen. Of all these metals, cobalt and iron catalysts have found widespread industrial application. Iron catalysts give higher selectivity toward lighter alkanes, and would be used for naphtha range products. Cobalt catalysts [328] provide higher selectivity toward the middle distillates and waxes. The synergy of catalyst, catalytic reactor and reaction conditions are the key issue [327] in obtaining high and stable yields of clean hydrocarbon fuels and waxes.

The performance of these catalysts is affected by numerous factors, one of which is the nature and structure of the support materials. Most studies on FT catalysts have been performed with the metals supported on silica, alumina, or titania [327–331]. For metal oxide supports, if the catalyst particles are very small (<4 nM), cobalt may react with the support to irreducible mixed compounds during either synthesis or catalysis [327,329,332,333].

The effect of catalyst particle size has unexpectedly proven to be very important. γ-Al_2O_3 with larger size catalyst particles has been the standard support being used for catalysts in FT synthesis [334–336]. Co on γ-Al_2O_3 has low selectivity, but good activity for longer chain alkanes. Recently, Eri et al. [328] obtained significantly higher C_5 + selectivity by using cobalt supported on low surface area α-Al_2O_3 (15 m^2/g). Though the reaction rate was lower than the γ-Al_2O_3 loaded Co.

Thermal effects during FT synthesis are another critical factor in reactor design, catalyst life and eventually selectivity and the propagation factor. The FT reaction is highly exothermic. The exothermicity combined with a high sensitivity of product selectivity to temperature constitute the main challenges in the design of FT reactors today. The alumina support has limited capability to rapidly transfer the heat to the environment, creating an instantaneous local spike on the catalyst particle. Complicated external cooling loops have to be employed in the most popular slurry bed reactors. The catalyst is suspended in the product slurry and recirculated between the reaction chamber and a cooling mechanism. The oil provides the large heat sink surrounding the ceramic supported catalyst. Despite this technique, local temperature spikes occur. The thermocouples monitor the fluid temperature, and cannot detect these local temperature spikes. High local temperatures make the metal particles mobile, which results in agglomeration of the particles, ultimately reducing the surface area [337].

The common factors of oxidation and coking from the reactants are also present. Exacerbated by the metal–oxide interaction, the water produced during the reaction can oxidize the metal [338–340]. Polymeric carbon was found on the catalyst particle as well as the alumina support by Moodley et al. [341]. The data was collected from an analysis of catalyst particles sampled from a running slurry bed reactor. Carbon growth was slow, but definitely a contributing factor for the long term deactivation of the catalyst.

The issues faced by the industry can, in part, be addressed by using GN as catalyst substrates. From a cursory look, I would like to discuss my thoughts and experience before further examining the published literature:

GN catalyst supports form SMSIs, yet there are no oxides in the substrate. the catalyst particles are flat and immobile.

- In the FT reaction, the exothermic reaction requires constant heat removal. The slurry reactors aim to rapidly transfer the heat to a liquid in which the catalyst particles are suspended, and consequently cooled with an external loop. Unfortunately, heat transfer from the metal catalyst particles to the liquid is via the support. The γ-Al_2O_3 support acts as an insulator, resulting in frequent temperature spikes localized at the metal catalyst particle. At the typical LTFT operating temperature of 250°C, the thermal conductivity of graphite is almost 4 times that of aluminum oxide. This property, combined with almost 100% of the surface area being external, helps carry the heat of reaction away from the catalyst much faster. This could mean a potential drop-in type of application for GN. One of the references cited above [337] compares the alumina substrate with silicon carbide. Silicon carbide has a very high thermal conductivity and is extremely hydrophilic. In the form of foams, can provide excellent heat conductance, rapid absorption/desorption of the produced water.

- SiC, especially β-SiC has proven to be a very good catalyst substrate for FT and many other exothermic reactions. In a patent application, SiCat [342] describes the improved stability of cobalt on SiC substrates, with mesopore structures of β-SiC. SiC supports are also discussed by Liu [343]. The mesopores overcome a major disadvantage with ceramic substrates. The formed alkanes (particularly waxes) sometimes do not have a path to exit the reaction zone rapidly enough with alumina supports, creating occlusion and restricting access of the reactants to the catalyst sites. Our group has worked with SiCat (France) to incorporate GN into the β-SiC substrate structure. We have incorporated GN on the surface of porous β-SiC discs to enhance performance. The results are encouraging. Again, probably due to the fact that the catalyst particles are not encased by any substrate, and products can diffuse out faster. The combination gives a heat sink as well as a low diffusional resistance catalyst substrate. They are strongly attached to the edges of the GN. Unfortunately, supply is limited, and mass manufacturing of the final product (β-SiC + GN) would require significant capex which catalyst manufacturers don't seem to be willing to invest in.

Back to published literature, GN have been investigated for use in the FT reaction with success. Yu used GN [344] as supports for Co FT catalysts. The activity and selectivity of the GN supported catalysts were studied at 210°C, 20 bar, and $H_2/CO = 2.1$, and compared with corresponding activity and selectivity for α-Al_2O_3 and γ-Al_2O_3 supported Co catalysts. The GN supported catalyst demonstrated both high activity and high selectivity to C_5+ hydrocarbons, with activity comparable with Co/γ-Al_2O_3 and selectivity comparable with Co/α-Al_2O_3. Bezemer [347] studied the potential of using GN as FT catalyst supports with cobalt loadings varying from 5 to 12 wt%. The activity at 1 bar syngas varied with increasing loading from 0.71 to 1.7×10^{-5} mol CO/gCo-s.

Stable activity of 225g_{CH_2}/lcat-h for 400 h was obtained at pressures of 28 + bar syngas. A C_5 + selectivity of 86 wt% was achieved, which is remarkably high for an unpromoted catalyst.

I discussed the effect of curvature dependence on the ability of CNTs to perform better in catalysis applications [304] and the behavior of the curved surface (hexagonal CNTs) as edge zones on GN. Using Fe catalysts on CNT walls, Menon et al. [307] report high conversion rates and low selectivities for methane and C_2 alkanes and olefins, while the bulk of the conversion and selectivities are directed toward the formation of naphta products (C_2—C_{11}).

GN catalyst supports form unique metal substrate interactions, yet there are no oxides in the substrate. The catalyst particles are flat and immobile. Zarubova et al. [345] have compared FT performance between platelet type and herring bone type nanofibers. In one set of samples, they synthesized both GN by growing them on macroscopic carbon fibers, similar to our approach of SiC. The other set included both type of GN used as powders. Cobalt was deposited and TOF (turnover frequency) selectivity and reaction rates recorded. TOF is the ratio between the weight of products and the percent of active catalyst sites (%Co × %dispersion) In this case, since the rate of reaction was measured, the TOF was calculated as the rate × MW cobalt. They report that the selectivity toward methane of the GN on carbon fibers was lower than that of GN powders. Between the two, platelet type had the better selectivities (CH_4 as well as C_5 +) than herring bone for both types of samples. The reaction rate was highest for the platelet sample. However, the TOF of the platelet GN-CNF was nearly half of all the other samples. The authors explain the difference in terms of access of much higher dispersion of cobalt on the platelet sites, making a smaller portion available to the reactants as active sites due to the occlusion resulting from the carbon fibers.

Bezemer et al. have multiple publications toward the prospect of using GN as supports for the Co in FT reactions [335,346,347]. I am sure there are many other interesting publications for this high potential application. Some others I have had the opportunity to peruse [348—350].

7.3.3 Hydrogenation reactions

Background: Hydrogenation is a reaction between a compound with unsaturated bonds and hydrogen (H_2). The process is commonly used to saturate the unsaturated bonds or for reduction reactions. Pairs of hydrogen atoms are added to a molecule by "syn" addition, such as in the reduction of double bonds in an alkene. The reaction requires very high temperatures without a catalyst.

The largest worldwide application of hydrogenation is for the processing of vegetable oils. As produced, vegetable oils are poly-unsaturated fatty acid derivatives. Hydrogenation (partial) reduces the number of carbon–carbon double bonds, giving the oils more commercially desired physical properties such higher melting points, longer shelf life, etc.

It is also a very important reaction in the oil industry. Hydrogenation is used to saturate alkenes and aromatics into saturated alkanes and cycloalkanes respectively. Similar to saturated vegetable oils, saturated fuels have longer storage life. Heavy residues in refining are hydrogenated by hydrocracking into diesel and naphtha.

It is a widely used reaction in the chemical industry too. It is the most investigated reaction for possible substitution of catalyst substrates by GN, both in the liquid and gas phases.

7.3.3.1 Selective hydrogenation of alkenes and α,β-unsaturated aldehydes

Hydrogenation studies were done with light alkenes such as ethylene, 1-butene and buta-1,3-diene with nickel catalysts supported on different types of GN, γ-alumina and activated carbon [123,303,351]. The authors report that the catalyst supported on GN allows higher conversion, compared to those obtained with γ-alumina and activated carbon supported systems. These results were relatively independent of catalyst particle size (6.4–8.1 nm for GN, 5.5 nM for activated carbon and 1.4 nM for alumina). This implies that catalytic hydrogenation might be a structure sensitive reaction. High resolution transmission electron microscopy (HRTEM) studies have given insights onto the metallic particles morphology; on GN supports, the deposited crystallites were found to adopt very thin, hexagonal morphologies exhibiting SMSI.

- Oxidized GN supported rhodium nanoparticles (1.1–2.2 nM) were used as a catalytic system to study the hydrogenation of cyclohexene [352]. These catalysts turned out to be extremely active even at low hydrogen pressure with low metal loadings (1% (w/w)) and low cyclohexene concentrations (1% v/v). The authors propose that the activity is almost independent of particle sizes of the metal and that other factors like the possible clustering of the support in the liquid phase and the influence of oxygen-containing species present on the surfaces of the support are responsible for the final results. A similar Rh/GN catalyst prepared under mild conditions, displayed comparable activity for the same reaction. No comparison was shown in this study against Raney Nickel, which is an economic alternative to the precious metal hydrogenation catalysts. When exotic metals are used for catalysis, the cost of the precious metals dominates, hence the ease and cost of recovery would be a factor in the commercial analysis. GN offer cost saving advantages in such applications. First, the catalyst would last much longer due to better dispersion and SMSI, second, when regeneration is required, the metal would be dispersed on the external surface area of the GN, and would be relatively easy to remove with an acidic wash without affecting the GN, rendering the GN suitable for reuse.

7.3.3.2 Hydrogenation of α,β-unsaturated molecules on GN

The hydrogenation of α,β-unsaturated molecules on GN supported catalysts has also been the object of several studies. Gas phase hydrogenation of crotonaldehyde to crotyl alcohol was conducted between 75°C and 150°C on a 5% (w/w) Ni catalyst supported on GN and γ-alumina [352,354]. Again, despite the large difference between particle sizes, (1.4 nM γ-alumina and 7 nM for GN), higher selectivity and activity were obtained on the GN supported catalysts. We can now attribute this to the fact that, for GN, nickel particles are located on the edge sites of the support. Thus it might be expected that between the two, different crystallographic faces of the metal will be exposed to the reactants. In addition, the possibility that electronic perturbations induced by the support could lead to different activities.

7.3.3.3 Cinnamaldehyde to hydrocinnamaldehyde

The liquid-phase hydrogenation of cinnamaldehyde to hydrocinnamaldehyde has been performed using a GN supported palladium catalyst with 50 M^2/g, and compared to catalyst using activated carbon with 1000 M^2/g surface area. The GN supported catalysts showed >90% selectivity compared to 40% selectivity exhibited by activated carbon supported catalysts and also showed much higher activity. The author explains the higher performance due to the absence of mass-transfer limitations on GN supports, compared to the microporous-activated carbon, and a peculiar graphitic carbon–palladium interaction may have favored some CO bond hydrogenation [353,355,358].

The same reaction was studied [359] with GN grown on β-SiC, a relatively new but versatile catalyst support, itself going through a cost reduction evolution process. The GN provided a high dispersion mechanism with a higher external surface area. According to the results, the hydrogenation rate on the Pd/GN/β-SiC composite catalyst was about twice of the one obtained on the Pd/β-SiC catalyst under similar reaction conditions.

7.3.3.4 Liquid phase hydrogenation of benzene

The liquid phase hydrogenation of benzene is an important industrial process because of its use for the production of cyclohexane. Cyclohexane is the main precursor for the production of nylon products. Therefore this reaction has very large market potential, if the costs of the intermediate products for nylon can be lowered even by a tiny percentage. Historically, cyclohexane used to be obtained by the direct fractional distillation of suitable crude petroleum refinery streams. Nowadays, the major portion of cyclohexane is obtained from the direct hydrogenation of benzene. Conventionally the reaction is carried out in vapor or mixed phase using a fixed bed reactor. The reactor temperature is between 180°C and 260°C. The activities of the Ni/GN, Ni/GN-α-Al_2O_3, and Ni/γ-Al_2O_3 catalysts were investigated in the liquid phase hydrogenation of benzene to cyclohexane. A 100% yield was obtained on the GN supported Ni after 7 h on line, whereas yields with Ni/α-Al_2O_3 were in the single digits [356,360]. The reason may be due to the intrinsic properties of the GN support. First, the GN have very effective specific surface. Second, GNs have metallic and semiconductive properties, which favor the electron transfer and spillover of hydrogen during the reaction.

7.3.3.5 Pyrolysis oils

A potential new application for small systems is in the hydrogenation/hydrotreating of pyrolysis oils. Pyrolysis is the process of thermally degrading long chain polymer molecules into smaller, less complex molecules through heat and pressure management. The process requires intense heat with short duration in the absence of oxygen. Pyrolysis is being used to convert waste products such as tires, plastics and municipal solid waste to useful energy products. The calorific value of this oil is not quite the same as fossil fuel due to the myriad of liquid compounds produced. The three major products that are produced during pyrolysis are oil, gas and char. Depending on the feedstock used, and operating conditions, pyrolysis oil composition can vary widely from one system to another, producing a mixture of aromatic compounds, oxygenated compounds, organic acids, tars and many more. There is much work going on to understand the nature and mechanisms of formation of this oil. Many activities are focused on investigating and developing processes to upgrade this oil by hydrogenation and hydrodeoxygenation [361–374]. Pyrolysis is one of the few processes that are inherently carbon-negative, if the feed is waste of any kind, and all of the products are utilized. For this reason, it will continue to grow as a market, especially the upgraded pyrolysis oil. This application is in the early stages of development and can provide a good start for GN use. As the market for small, distributed pyrolysis systems grows, lowest cost processing will be required by the operators, and the manufacturing capacity of GN-based catalysts can also grow with the budding industry. There would be no baseline or paradigm to change. Manufacturers wouldn't have to take any risks of switching from an established process.

A serious issue in catalytic hydrodeoxygenation appears to be catalyst deactivation due to thermally induced polymerization and the formation of coke [375–377]. Coke deposition on the catalyst has been shown to occur particularly with catalysts based on alumina supports [368]. GN were also proven to be superior catalyst supports in other liquid phase hydrogenation reactions, as well as the hydrogenation of chloronitrobenzenes, one of the possible compounds produced by pyrolysis in the presence of chlorinated plastics such as PVC [378].

7.3.3.6 Vegetable oils to renewable drop in diesel

Another interesting application is the decarboxylation or hydrodeoxygenation of vegetable oils to produce alkanes in the C_{16}–C_{22} range, directly compatible with fossil diesel fuel [379–384]. It may not be viable for virgin oils, but could produce a much higher value product than biodiesel if one was to use waste oil from oil processing and industrial processes which use vegetable oils for single or double use. There is, however, a total mass loss of 11% due to the loss of the oxygen in deoxygenation and an additional 3–4% in decarboxylation. Palm oil, for example will have a mixture of fatty acids from C_{16} to C_{22} with an average molecular weight of 280. It is a saturated oil, but has 2 oxygenated bonds. Loss of oxygen lowers the molecular weight by 32. Another cost to consider is that of hydrogen. High partial pressures of hydrogen are required for the reaction. In the commercial sense, these two factors raise the cost of the final product. However, methyl esters of fatty acids normally called Biodiesel demand a premium in the fuel markets, but the esters still retain the double bonds and oxygen, resulting in a very high cloud point (temperature at which the biodiesel turns cloudy due to formation of solids) and pour point (temperature at which biodiesel starts/stops pouring as a liquid). Gelling at high temperatures plugs the diesel filters in cold climates. Hence the biodiesel is always blended with the fossil diesel in small weight fractions (5–15%). A drop in diesel (direct substitute, also known as renewable diesel or second generation diesel) will have identical properties as regular #2 diesel, and demands an even higher premium. In the US, RINs (Renewable Identification Numbers) for renewable diesel provide credits that can be traded for substantial revenue addition. In California, the revenues double. The higher revenues more than offset the loss of mass I just mentioned. For the hydrogen component, one could get creative and place the GN and hydrodeoxygenation plants in the same complex. If catalytic methane cracking is used for GN production, the by-product hydrogen could partially offset the cost of hydrogen. The use of GN as substrates has been investigated for decarboxylation and hydrodeoxygenation of vegetable oils [381,384,385,386,387,388]. The substrate used presently is activated carbon with a precious metal like paladium. Although this is a very interesting application, I place quite a few caveats for the commercial success. Waste oil is sought after aggressively by the biodiesel manufacturers. If it has a high percentage of fatty acids from overuse, it can be converted into methyl esters (biodiesel) with a simple esterification reaction using sulfuric acid as a homogeneous catalyst. If there has been very little oxidation and the free fatty acid content (FFA) is low, an even simpler process, transesterification can convert the oil to methyl esters.

7.3.4 Dehydrogenation

Dehydrogenation can be achieved by direct removal of hydrogen from alkanes. Catalytic dehydrogenation of alkanes is an endothermic reaction, which occurs with an increase in the number of moles and can be represented by the expression

$$\begin{array}{ccccccc} (CO_2)_g & & & & & (CO)_g & \\ \updownarrow & & & & & \downarrow & \\ [CO_x]_{ads} & \rightleftarrows & [(Co_x)_i]_{ads} & \rightarrow \cdots \rightarrow & & [C{-}C{-}C{-}C{-}C] & \rightarrow \\ \updownarrow & & & & & & \\ (CO)_g & & & & & & \end{array}$$

The C–C bond strength (about 246 kJ/mol) is lower than that of the C–H bond (about 363 kJ/mol) hence direct thermal cracking would favor the undesired direction. The strong C–H bond is a closed-shell σ orbital and would be activated by oxide or metal catalysts. Oxides would be able to affect the stronger C–H bond, by forming O–H bonds (with strengths close to C–H bonds), due to hydrogen abstraction. The metals by themselves would not be able to pull hydrogen because the bonds between the metal and hydrogen are weaker than the C–H bond. The reaction proceeds by utilizing the combined energy of these two types of bonds. The proposed reaction mechanism is via a transition state, which can be described as a metal atom inserting into the C–H bond. The C–H bond bridges across the metal atom until the C–H bond breaks, followed by the formation of the corresponding M–H and M–C bonds.

The dehydrogenation of ethyl benzene to styrene is also an important reaction in the industrial world. Almost all the ethyl benzene produced by the Friedel Crafts reaction is consumed for this process. The product, styrene monomer is used to make polystyrene, a widely used plastic and to make other copolymers. Styrene itself is a liquid, easy and safe to handle. The vinyl group makes styrene easy to polymerize and copolymerize.

The incumbent catalytic process for industrial production of styrene is achieved by the catalytic dehydrogenation of ethylbenzene. In the manufacturing process flow, the unreacted ethyl benzene is separated from styrene by fractional distillation and recycled. Fresh ethylbenzene is then mixed with the recycle stream and vaporized. Steam is added before feeding the effluent into each of a series of reactors. The process is highly endothermic and is carried out in the vapor phase over a solid catalyst. Iron oxides with alkali and chromite promoters are typical catalyst formulations used. Steam provides heat, prevents excessive coking, helps shift equilibrium of the reversible reaction toward the products, and cleans any coke that does form on the surface of the catalyst. The reactors are run adiabatically. Conversion (EB) and selectivity (ST) are commonly known to be in the 60–70% and 85–95% range respectively. Typical operating pressures are low because the primary reaction produces 2 moles of product for each mole of feed.

$$C_6H_5CH_2CH_3 = C_6H_5CH = CH_2 + H_2$$

Operating temperature is typically between 600°C and 650°C.

Oxidative dehydrogenation (ODH) is a new alternate route to accomplish many dehydrogenation reactions. ODH is an exothermic process compared to the highly endothermic route currently utilized in industry as explained above. Ostensibly then, ODH has the potential to reduce costs. Some of the benefits include less coking, eliminating the high temperature furnace and obtaining higher yields of olefins. ODH can be applied to some very important industrial small chain alkane to olefin conversions. Ethyl benzene to styrene is a promising and potentially large volume application.

$$C_6H_5CH_2CH_3 + \frac{1}{2}O_2 = C_6H_5CH = CH_2 + H_2O$$

The ODH conversion of ethyl benzene to styrene in the presence of oxygen on various pristine and oxidized GN surfaces has been studied extensively [389,390]. Von der Fakultät [395] compared the performance of the ODH reaction of ethylbenzene to styrene using carbon black, pristine graphite, GN, and MWNTs at 550°C. The data is compiled in the table below for easy reference. The author does not give the O_2/EB mole ratio in the feed. Specific values reported below are the relevant values divided by the specific surface area. Specific surface area is defined as the surface area of the catalyst per gram × the weight of catalyst used (Table 7.6).

Xu et al. [309] have reported different values from Von der Fakulat, with a reported O_2/EB mole ratio of 1.1 They report a comparison between a commercial catalyst and GN in very dilute reactant flows. (98% inert gas, He) (Table 7.7). I will expand on this data below.

Zhao et al. reported more useful data [392]. They performed the experiments in a lower temperature regime (350–400°C). Lower temperatures of operation are one of the claims of the proponents of ODH over conventional catalytic processes. They compared the performance of different types of GN, identified by the catalysts used to synthesize them, and the

Table 7.6 Experimental results oxidative de-hydrogenation of ethyl benzene to styrene [391–394]

	Conversion (%)	Specific selectivity	Specific yield	Benzene (%)	Toluene (%)	CO (%)	CO_2 (%)	ST (%)
GN	64	4.97	4.67	4.3	1.7	1.9	10.5	42.3
MWNT	55	0.43	0.23	0.7	0.12	2.4	22.10	28.71
Graphite	52	2.94	2.93		5		4–6	N/A
Arc discharge MWNT	70	5.35	4.15	1.04	0.54	9.85	11.70	56

Table 7.7 ODH of ethyl benzene to styrene [309]

	GN (%)	Commercial catalyst (%)
$Selectivity_{ST}$	93.9	73.3
$Conversion_{EB}$	39.8	13.9
$Yield_{ST}$	37.0	9.9

Table 7.8 Performance of various catalysts for ODH of ethyl benzene to styrene

No.	Catalyst	Conversion (EB)%	Selectivity (ST)%	Yield ST (%)
1	Ni/Al_2O_3	26.45	77.23	20.42
2	Fe/Al_2O_3	58.12	84.96	49.38
3	Ni_2-Fe/Al_2O_3	48.47	80.12	38.83
4	Fe powder	42.25	70.56	30.98
5	1-HCl treated	58.83	85.06	50.01
6	2-HCl treated	47.27	82.04	38.75
7	1-Heat treated	40.23	86.67	34.84
8	2-Heat treated	35.12	80.86	28.83

post synthesis treatments they were subjected to. The Fe/Al_2O_3 catalyst at 600°C with CO and H_2 gave the best performance. Acid washing of the fibers did not improve the performance but heat treatment drastically reduced the performance, providing clear evidence that the edge sites need to be in the planar shape for maximum oxidation to support the reaction. See Table 7.8.

Economics: Is the ODH process ready for the real world? Unfortunately, the scientific data generators are often either oblivious of the industrial realities or feel obligated to report the positive side of their results only, as the analysis on ethylene oxide showed us. The research is valuable for progress, but the identification of hurdles is just as important as claiming that the results show the reaction is suitable for industrial use. Let us look at some aspects:

1. To my knowledge, the typical selectivity of the steam-based direct dehydrogenation system used predominantly by industry is very high, i.e., much more than 95%. However, inherently this process can tolerate lower selectivity, without much of an economic impact, because benzene and toluene are the by-products that can be recovered for sale/reuse. On the other hand, selectivity is critical for ODH. Lower selectivity will naturally result in products of combustion of ethylbenzene, and consequently wasted. For example, consider the data from Zhao et al. [392], they have presented a comprehensive comparison of

as-produced and treated nanofibers for use as catalysts in the ODH of ethylbenzene to styrene. The optimum result ($\sim$60% conversion, 85% styrene selectivity, 12% CO_2 selectivity) is obtained by acid-treated herringbone GN. Though the edge sites are shown to have higher activity, this comparison only serves to show the relative performance of the different nanofibers. Let us take a closer look: Assuming the optimum result is identical to the numbers achieved by the incumbent process, and using data from Zhao, Fig. 7.10 A shows the mass balance with ODH and Fig. 7.10B shows the mass balance of direct dehydrogenation used by industry today. The loss of feed as CO_2 is significant in Fig. 7.10 A. In the real world, the selectivities are much higher than assumed here, and the numbers end up looking even less attractive. To be clear, Zhao et al. do not claim their work could be applied industrially to reduce costs of production of styrene. But the data

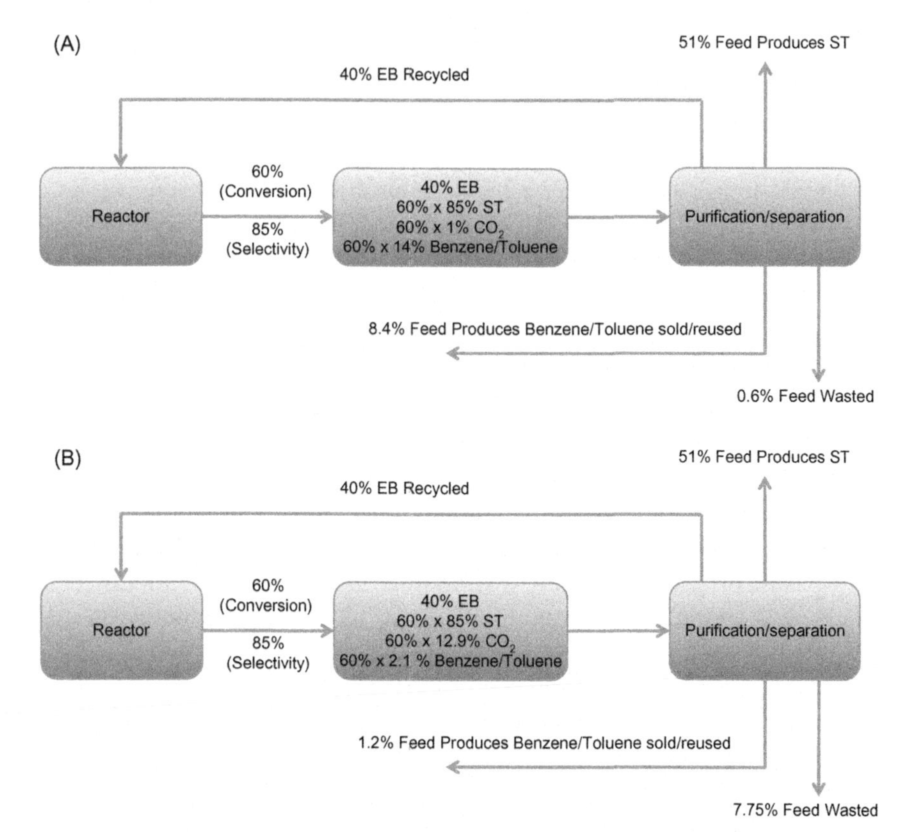

Figure 7.10 (A) Incumbent Industrial Process for EB to Styrene. (B) ODH process for EB to Styrene – Data from [392].

should not be misconstrued as that consequence being an obvious reality. However, Xu et al. do claim ODH can replace direct dehydrogenation of ethyl benzene [309] due to the energy savings afforded by the exothermic ODH reaction. The conversion data presented does not seem to have been derived using the knowledge mentioned earlier in the document about the coke formation being a major contributor of catalytic activity. The commercial catalysts were run for only 6 hours, while coke formation, and consequently the catalytic activity of the "commercial" catalysts increases over time. The conversion rate of 4% for the commercial catalyst is unrealistically low under the operating conditions. The data is not sufficient to determine how long it took the alumina catalyst to achieve the equilibrium temperature of 547°C. The experimental section does not specify if the reactor and commercial catalyst were preheated or not, and what the exit temperature of the helium gas was. Based on T_2-T_1 of the He, one could calculate the total heat transfer that occurred to the alumina catalyst per unit of time and determine the length of time it took to reach the reaction temperature. At 100% heat transfer efficiency, the catalyst would take 25 minutes to reach the reaction temperature. At 7200 per hour GHSV (calc.), this would be highly improbable. Without long derivations, it would be reasonable to assume that on the average the heat transfer efficiency would be closer to 10–30% as the differential temperature keeps getting smaller over time. At this rate, the catalyst itself would take approximately 2–4 hours to heat up to the reaction temperature. Therefore the effect of a relatively cooler catalyst surface temperature on the reactants is important to study and analyze.

2. O_2:ethyl benzene ratio. Another issue with publications is the oxygen to ethyl benzene ratio. The ratio is often not specified in research papers. High ratios will cause combustion as well as an explosion hazard. This consequence is avoided in research efforts by performing the experiments with extremely low volume fractions of O_2 and EB. For example, referring to Ref. [309] again, all data has been collected with a 98% He volumetric fraction flow, which is completely unrealistic in an industrial environment.

If we now take a pragmatic look at the ODH route to EB–ST, at the least, there is much work required to meet the performance of the incumbent process. The reason I went through this analysis again is to demonstrate the implied results and how the end user sees the results.

Having said that, the energy savings offered by the ODH route are very impressive. The exothermic nature of the reaction would first present the same challenge of heat dissipation, and the catalyst sintering that follows for the conventional catalyst substrates, as we discussed in Section 7.3.2. Hence the GN would shine there. In my opinion, the task here is to develop a catalyst using GN as a substrate that inhibits the combustion route, and hence the waste of EB as CO_2. To try the process with GN, we must achieve similar selectivities and conversions with very high GHSV values, such as 30,000 per hour. The reaction kinetics and models for such an endeavor will need to be developed simultaneously.

Opinion—Summary: The ODH reaction has many positives, but I was not able to find references which compared the ODH route to the incumbent process in an apples-to-apples manner. My detailed analysis tells me that given the potential, it is a worthwhile pursuit, but claims for commercial readiness are premature.

Calculations for Ref. [309]:	
Heat capacity of He:	5.19 J/g-K = 0.0052 J/mg-K
Density of He at constant pressure:	0.164 mg/mL
He flow rate:	9.18 mL/min
Mass He per minute:	9.18 mL/min × 0.164 mg/mL = 1.5 mg/min
Heat available from He:	1.51 mg/min × 0.0052 J/mg-K × 550°K = 4.32 J/min
Heat required by catalyst:	
Catalyst weight:	21 mg
Fe weight at 20%:	(0.2/1.2) × 21 = 3.5 mg
Heat capacity of Fe:	0.000444 J/mg-K
Heat required by Fe:	3.5 mg × 0.000444 J/mg-K × 500°K = 0.78 J
Alumina weight:	21 mg − 3.5 = 17.5 mg
Heat capacity alumina:	89.7 J/mol-K = 0.012 J/mg-K
Heat by required by alumina:	0.012 J/mg-K × 17.5 mg × 500°K = 105 J
At 100% heat transfer efficiency:	
Catalyst heating time:	(105 + 0.78) J × 4.32 J/min ~ 25 minutes
At 50% heat transfer efficiency,	
Catalyst heating time:	~ 50 minutes
At 10% heat transfer efficiency,	
Catalyst heating time:	250 minutes or 4+ hours
GHSV calculation:	
Density of alumina:	3.95 g/cc
Volume of catalyst:	3.95 g/cc ÷ 0.021 g = 0.083 cc = 0.083 mL
Gas flow:	10 mL/min = 600 mL/h ~ 7200 per hour

Other interesting research on this reaction using GN includes the following:

1. Weinstein [393] reports using GN as supports for the ODH of ethanol to acetaldehyde and ethyl acetate.
2. Sui et al. [394] report the successful ODH of propane on GN-based phosphoric oxides at lower temperatures of 500°C to replace non-catalytic and non-oxidative processes with vanadium and molybdenum oxides. A 39.63% propene selectivity was reached at a 42.07% propane conversion over GN-supported 5 wt% phosphoric oxide catalysts. Given the higher active specific surface area of GN, a vast improvement in this performance can be expected.

7.3.5 Fuel cell electrodes

Before there was graphene, there were fuel cells. (PEMFCs) Proton exchange membrane fuel cells, drew a great deal of attention in both fundamental and application research a couple of decades ago, and the activity still continues, though on a much smaller scale. PEMFCs have unique and favorable advantages over chemical batteries in terms of high efficiency, high energy density, and zero or low emission. Significant progress has been made in the research and development of PEM fuel cells in the last 50 years. Stack power density has increased, portability of fuel cells in vehicles has improved and conversion efficiencies have gone up. However, these improvements have failed to bring about the reduction in costs to justify use of fuel cells. The costs of producing hydrogen, the cost of catalysts and the cost of bipolar plates in PEM fuel cells make the package unattractive for large-scale use as once projected, especially with the rapid improvements in the battery world, notably the Li-ion batteries. Nevertheless, the technology is extremely well suited for some applications. A notable one is the standby power units or uninterrupted power supply units for cellular towers, and large critical power consuming operations. With relevance to GN, one of the cost reduction paths proposed by the nanomaterial manufacturers was the possibility of reducing the catalyst loading on the anode of the fuel cell, when the catalyst used were almost exclusively precious metals.

I have mentioned catalyst sintering as an issue before. The very same quality makes the GN a lower cost alternative. Since the catalyst cannot be replaced easily in a fuel cell, the sintering effect is accounted for upfront and more than the required amount of catalyst is deposited. Differences in catalyst morphology is also said to affect the performance. Conventional catalyst particles deposit on the carbon black as globules, with very small active surface area. All the benefits of the edge surface area in GN apply in this application. Lastly, the impurities in carbon black contribute to premature poisoning of the catalyst, unlike GN, which are highly crystalline materials. The GN electrodes have displayed higher activities than the ones using Vulcan carbon [395].

Fuel cells seem to have been the red herring of the last century. They have so far only been practical for niche applications. Nevertheless, GN can be beneficial to help improve the cost to benefit ratio of PEMFCs. They can contribute to this effort in ways beyond catalysis. A couple of my thoughts:

1. PEMFCs typically have a gas diffusion layer (GDL) between the anode and the PEM. The proton exchange membranes typically do not have very high surface smoothness. The gas diffusion layer (as the name implies), allows more complete transport of hydrogen to the surface of the PEM by intimate contact. If GN are used as catalyst supports, the high surface area of platelet GN from the edges, would provide the close contact with the uneven surface of the membrane, and possibly eliminate the GDL.
2. Due to their highly crystalline nature and low cost, GN could be a viable raw material for manufacture of the graphite bi polar plates in the PEMFCs.
3. On the cathode side, a hydrophobic catalyst support would improve life. Water is produced at the cathode of a fuel cell, and carbon is less hydrophobic than graphite.

7.3.5.1 Summary

Catalysis is a very complex field, and much knowledge is yet to be gained for higher efficiencies and lower costs. Most catalysts are metallic, and given the unique interaction of the edge sites in GN with metals, especially transition metals, there is tremendous scope for improvement in this field. We have only just started understanding quantum mechanical effects in chemical interactions. Electronic perturbations by the energy density on the GN edges or by external stimulus can perhaps change the way catalysis is looked at in the future.

From the examples we studied in this section it seems that GN can possibly provide enough improvements to justify industrial scale involvement. Data reported so far only "points" toward higher performance. There should be a concerted effort to investigate further. My personal experience can only vouch for use of GN in FT, pyro oil upgrading and decarboxylation reactions. As we have discussed earlier, the organic synthesis applications will probably be a long time before adoption.

7.4 WATER & WASTEWATER TREATMENT

Water has become the most valuable commodity in the world today. Unprecedented amounts are being pumped from the ground for agriculture and dumped into the ocean, contributing to the rise of the levels of oceans, which then has a negative impact on the water tables in coastal areas. A rise in the water tables saturates the soils and water is discharged without any use derived. Yoshida et al. [396] claim that we are extracting groundwater at an unsustainable rate. Recycling this finite resource is the only solution for sustainability of the growing population. Countries like Singapore have taken major steps to recycle their water. This step may have been to reduce dependence on neighbor country Malaysia, which supplies almost all the water to the tiny city-state. Nevertheless, water is being recycled. The desalination industry is growing at a rate of >14% compounded annual growth rate (CAGR). In most instances, desalination plants provide potable water at the expense of great amounts of energy, while the same water, on its way out is minimally treated and mostly discharged.

Reverse osmosis, which is the most popular desalination technology, has come a long way since I first got involved in it, in the early 1980s. The membrane costs have come down substantially, membrane life has gone up, and power consumption has dropped significantly. Yet, it remains beyond the reach of most developing nations. The biggest bang for the buck for sustainability would be to cooperate with large volume generators of wastewater and help them recycle their wastewater. Industrial wastewater and municipal wastewater treatment plant effluents are excellent targets.

7.4.1 Membranes

Theoretically, nanomaterials have the potential to improve the performance of membranes of various types. Perhaps there is a future of carbonaceous nanomaterials in the field of specialty membranes, though it will have to wait till we have overcome all the hurdles described before. Nevertheless, we can narrow down the possibilities, starting with theory.

Molecular Dynamic Simulations and models [397,398] indicate orders of magnitude increases in flux of water through a GO enhanced membrane or aligned nanotube openings. In practical experiments, GO composite nano-filtration membranes have been fabricated. Nan et al. [399] write about a nano-filtration membrane for commodity applications like water softening. They achieved a flux of 4.2 L/h M^2 atm. However, commercially available nano filtration membranes already boast a flux of $\sim$5.6−5.8 L/h M^2 atm [400]. Nano filtration membranes by nature of the manufacturing process have a surface charge and tend to repel divalent anions. However, they are not immune from the limitations of reverse osmosis membranes, such as biological fouling and osmotic pressure. One of the factors in membrane operations is the minimum surface velocity to avoid fouling. Therefore, there will always be a minimum flow rate per membrane. A multistage membrane system will still need a minimum flow rate in the final housing. Combining typical recovery rates (permeate flow rate/feed flow rate) per stage and osmotic pressure limitations, the water-softening application [397,398] doesn't come close to the recovery rate of ion exchange water softeners. Nanofiltration membranes, are, for sure, good tools for specialty separations. Sun et al. [401] talk about an ion permeation membrane using anionic GO and cationic Co−Al (or Mg−Al) layered double hydroxide nanosheet superlattice units for high selectivity charge-guided ion transport through the membrane with a concentration gradient between the salt and deionized water. One of their configurations shows a 5 times higher flux of monovalent ions than multivalent ions. Jia et al. [402] used dicarboxylic acids and diamines to bond basal planes and edges of a GO membrane, and investigated the relationship of metal ion flux with hydration level of the ions and GO. They found the hydration level of the GO influenced the flux for single salt solutions, while the diameter of the hydrated ion influenced the flux of the ion migration. GO membranes and papers also have the issue of mechanical strength. The hydrophilic functional groups in GO make them unstable in aqueous solutions. The mechanical strength of membranes is a major hurdle for higher flux and porosity. Improvements in the strength of GO membranes by crosslinking with various compounds have been reported. Park et al. [403] achieved up to 200% increase in mechanical stiffness and approximately 50% increase in fracture strength of GO paper with just 1% doping with alkaline earth divalent ions. Borate crosslinking achieved a storage modulus of 127 GPa [404], Tian et al. [405] with polymerized dopamine (PDA) and poly-etherimide (PEI) crosslinking, Jia et al. [406] by esterification reactions, using dicarboxylic acids, diols or polyols as the crosslinker to achieve 1560% increase in the elastic modulus.

The cost of making GO is prohibitively high to attempt to commercialize the above theoretical achievements, and the effort to enhance performance of an enhancement material is counterintutive to my simple brain. However, existing polymeric membranes could use some help from nanomaterials to increase their strength and maybe augment the porosity a little. The more porous you make polymeric membranes, the weaker they get, and more support they require. Almost all polymeric membranes today have a support layer attached to them.

The most interesting and practical field in membrane separations currently is what is termed Forward Osmosis. This nomenclature is a bit confusing, because what it actually defines is Osmosis. But since reverse osmosis is such an established process, where some people simply use Osmosis as a short form, the commercial pioneers at Hydration Technologies, Inc. (Robert Salter) coined the term Forward Osmosis. This is a new and upcoming field, and I find that most people are unfamiliar with the details and idiosyncrasies of this technology. FO is a process we are aggressively piloting for treatment of wastewater that has too complex a chemistry for conventional treatment, such as the wastewater expelled from our pyrolysis and gasification systems. I will share a brief narrative of our efforts in this field with GN-enhanced membranes, but largely, I will spend some time in explaining the technology, the current applications, the hurdles and possible role of GN.

As can be imagined, in osmosis, the water molecules flow from a body of low osmotic pressure to that of higher osmotic pressure. Nature displays this phenomenon in the water uptake mechanism of the plant roots and the deltas where ocean and rivers meet. FO can be used to extract pure water from an aqueous solution using a "draw" solution that has a higher osmotic pressure (higher affinity for H_2O). It follows that FO does not require any hydraulic pressure like reverse osmosis does (an explanation of how reverse osmosis works is not relevant to this discussion. The reader is referred to copious amounts of literature in the public domain). In FO, without applied hydraulic pressure, there is minimum fouling of any kind on the membrane surface, given the engineering of the established flow regimes is done correctly.

FO turns out to be a wonderful technique when the diluted draw solution can be a useful product. But these applications are rare and specialized. HTI has a very practical and interesting concept. They use polysaccharides such as concentrated syrups of different flavors for drinks as draw solutions. The diluted draw solution ends up being a drink with all electrolytes included. Perfect for emergencies. For large-scale applications, though, where purified water is required, an economic draw solution with a suitable osmotic pressure has been a major challenge. Two drawbacks of the current approach of using liquid draw solutions are:

1. There is always some diffusion of monovalent ions from the draw solution to the feed solution (back diffusion). This limits the type of compounds one can utilize. Back diffusion could contaminate the feed solution and make the concentrated unsuitable for discharge, unless the goal is to evaporate the concentrated feed solution and have no discharge.
2. High salt content liquid draw solutions end up creating a concentration polarization at the liquid/liquid interface within the structure of the membrane, dropping the flux rate. The FO process must therefore be dynamic, making the membrane modules complicated with two solutions flowing in opposite directions along the membrane interface.

To overcome these shortcomings, much work has been done on the practicality of use of solid draw agents such as reversible hydrogels [407–411]. Hydrogels constitute a group of polymeric materials which are soluble in water, and when cross-linked, take a solid form. The hygroscopic nature and stretching of the crosslinking bonds in these polymers renders them capable of holding large amounts of water in their 3-D networks. Hydrogels function in the critical phase region, and the water interaction is actually acquiescence. They react with the water molecule in the vapor phase as well as the liquid phase. But upon collapse, the water is dispelled as a liquid. This quality has been exploited for dehumidification, I have mentioned it earlier [236]. As mentioned in the polymer section, reversible hydrogels can be designed to be responsive to many different external stimuli. There are 8 different types of hydrogels classified according to the external stimulus that either activates them or initiates reversal. Hydrogels are used extensively in drug delivery, with pH as the external stimulus, for example in time-release drugs. Temperature sensitive hydrogels such as poly (*N*-isopropylacrylamide) collapse at their lower critical solution temperature (LCST). Thermally reversible hydrogels such as copolymers of poly(ethylene oxide) and (PPO) are important to our discussion of desalination. More information about other types of hydrogels is easily available elsewhere.

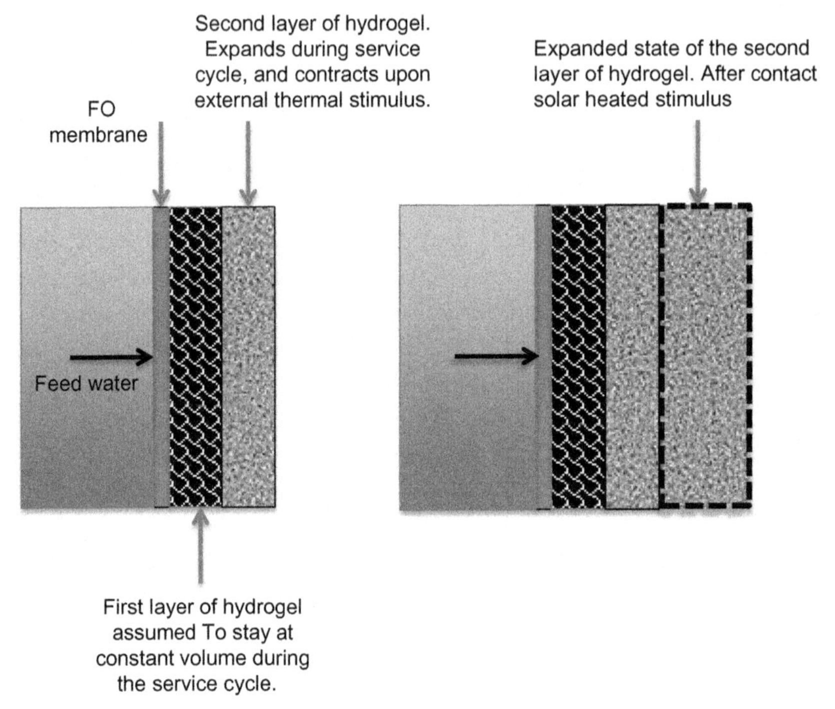

Figure 7.11 Double-layered hydrogel FO membrane composite [412].

For desalination, the hygroscopic nature of a hydrogel can generate a very high water activity force on the product side of an osmosis membrane, which we can call osmotic pressure. Cellulose triacetate (CTA) is a semi permeable polymer that can act as a barrier to charged ions and allow water molecules to pass through. As such, a combination of these two polymers would create an efficient FO system. Using hydrogels attached to CTA to affect osmosis has been shown. Razmjou et al. [412] demonstrated an in-situ forward osmosis system that prevents the layer adjacent to the semipermeable membrane from swelling. A double-layer hydrogel with a copolymer of sodium acrylate and *N*-isopropylacrylamide, and *N*-isopropylacrylamide was used as hydration layer and thermally reversible layers respectively. Solar-heated hot water is contacted with the reversible gel to bring the temperature above the LCST, and recover pure water. The authors do not report any long term results from this configuration (Fig. 7.11).

Our experience at the lab for long-term testing with wastewaters using a combination of FO/hydrogel method did not start favorably. The longevity and strength of the hydrogels was the dominant issue. We could design with the reversible gel grafted to the HTI membrane or the CTA coated Solupor® membrane. The strength of the hydrogels dropped over very short periods of time due to the cyclic nature of their function. By reason of diffusion ease and slow kinetics of hydrogels, for realistic flux rates and the regeneration times, the hydrogel layer has to be very thin, hence the mechanical weakness experienced. Using the dual layers outlined by Razmjou would still require a constant swelling/deswelling activity inside the first hydrating layer. It would be very interesting to do long term studies of >1000 cycles with Razmjou's concept. In our work, to overcome this issue, we incorporated 0.6% by weight of as-produced GN into poly(sodium acrylate)–poly(*N*-isopropylacrylamide) (PSA-NIPAM) hydrogels to improve the strength and cyclic life. Zeng et al. [413] have reported the same technique using (you guessed it) r-GO. They obtained higher fluxes at loadings of 1.2% r-GO. Our objective for using GN was mechanical strength rather than the swelling ratio. The hydrogel lasted for our entire standard (500 cycle) test. In our tests, the GN addition did not enhance the flux rate or the swelling ratio.

The world's water crisis is not going away, and since our largest source of water is seawater, it behooves us to find solutions sooner rather than later. We have the tools in nanomaterials, but we need to get out of the impressing stage and enter the practical stage. For desalination applications, the FO/hydrogel combination offers many advantages. Obviously, it would consume negligible energy, just the pumping for transport of water. Most of the process energy would be delivered by thermal solar energy, or very low level electronic perturbation as an external impulse to the reversible hydrogel [236]. A large market (and unfortunately more in need) is the thousands of island communities around the world. FO allows us to desalinate seawater without the use of chemicals. In reverse osmosis systems, chemicals added for prevention of fouling end up in the waste stream with the concentrate of the system. This waste is not allowed to be discharged to the ocean in areas where ecology is taken seriously. The high salinity and chemicals are detrimental to the tropical waters, where most populated islands are located. The local waters could probably have the ability to dilute slight increases in salinity, which would be the result of low recovery rates. For example, a seawater reverse osmosis system working on 3.5% salt content (35,000 ppm) and using ultrafiltration as a pretreatment, could possibly achieve conversion at 45–50% of the seawater feed. The economics look very good, but the resulting salinity in the wastewater would be twice as much as the feed (~70,000 ppm). By nature of having to use hydraulic pressure (1000 psi for seawater desalination), reverse osmosis systems are forced to control the economics by extracting as much fresh water as possible from the feedwater. The highest component of operating a reverse osmosis plant is the energy input.

FO/hyrdrogel systems have negligible power consumption, and capital costs are projected to be much lower, since high pressure components (membrane vessels, pumps) are not required. This scenario can provide the economics for very low recovery operation, and consequently, keeping the wastewater salinity increase to a minimum. A 10% recovery would increase the salinity by ~10–11%, that is, 35,000 ppm seawater would exit the system at ~39,000 ppm, requiring a 2:1 ratio of seawater to wastewater to bring the salinity back to the seawater levels.

Another process in polymeric membranes that is making a comeback and progressing at a rapid rate is membrane-based distillation, or sometimes called pervaporation. The process involves the use of extremely hydrophobic membranes that allow vapor to pass through but repel water. The mass transfer through the membrane is achieved through a vapor pressure differential between the two sides. The feed water side is maintained at a slightly elevated temperature (~70°C), (mostly achievable by waste heat or solar thermal hot water heaters), and the pure water side is maintained at a lower vapor pressure, either by a slight vacuum or cold purified water loop. The process is used for many liquid–liquid separations in industry, but the most active field is desalination. Banat et al. [414], Alklaibi et al. [415] and Hsu et al. [416] have looked at desalination. There are possibly many more publications out there which I have not had the good fortune to read. In addition, many investigations into solar thermal-based membrane distillation have been done by Galvez et al. [417], Hassan et al. [418], Ding et al. [419], Joachim et al. [420] and Walton et al. [421]. Water extraction from wastewater deserves a mention. Qu et al. [422] and Diez [423] report concentrating brine from reverse osmosis system reject. Gryta et al. [424,425] did work on concentration oily water, Calabro et al. [426] with textile industry wastewater, and Boi et al. [427] looked at wastewaters with volatile compounds.

The membrane distillation technology is viable on a technical level, but current average flux of 4–5 L/h-M^2 is not sufficient to compete with other modes of desalination. I have read research/academic papers on GO membrane fabrication for pervaporation [428–433]. In addition, our group has this application in the cross hairs. Clearly, the work is in progress for enhancement of these membranes. A much higher level of crosspollination between the polymeric world and the nano materials world could probably solve the issues critical to the success of the technology. From my experience with this technology, the membranes currently have some issues that can possibly be solved with GN enhancement:

1. More hydrophobicity would be welcome. The polymeric membranes in the market are not completely hydrophobic. Some water permeates through the fibers and forms a boundary layer. When evaporation takes place, the vapor carries with it the latent heat to the pure water side, cooling this boundary layer and changing it's vapor pressure thus reducing the evaporation rate We may view this as a similar problem to that of FO membranes, but in this case, we would see this phenomenon as creating a vapor pressure polarization effect.
2. Unless the feed water is heated with waste heat or solar heat, the energy lost from the latent heat carryover and conduction through the membrane, affects the economics significantly. The systems therefore need to improve upon thermal energy management.
3. To achieve better performance, membranes should be as thin as possible and have a lower thermal conductivity coefficient. For current polymeric membranes, if the thickness was reduced any further, the integrity of the membrane would be compromised at the elevated temperatures of the feed stream (typically 70°C). It would be beneficial to incorporate GN in some fashion to achieve improved strength and hydrophobicity.

We are currently working on this application (without any enhancements) with the Solupor® membrane (Lydall). The system is integrated with multiple passive solar distillation units. The membrane is made from ultra-high-molecular weight polyethylene and exhibits the highest strength-to-thickness ratio among the other membranes we have looked at (PTFE, PES, and PP). Fig. 7.12 depicts the current scheme installed in a remote village. The membrane system yields ~4–5 L/h-M^2 additional water from the hot concentrate of the solar stills, supplementing the distilled water volume.

In summary, there is much research going on in the field of GO and GO-enhanced membranes. Commercial applications on the horizon are few, but that could presumably be due to the cost of GO. Forward osmosis with the use of GN thermo sensitive reversible hydrogels can be efficient desalination systems. Solar thermal water heating devices would be sufficient to supply the energy required for this process, making safe drinking water accessible to the neediest of nations.

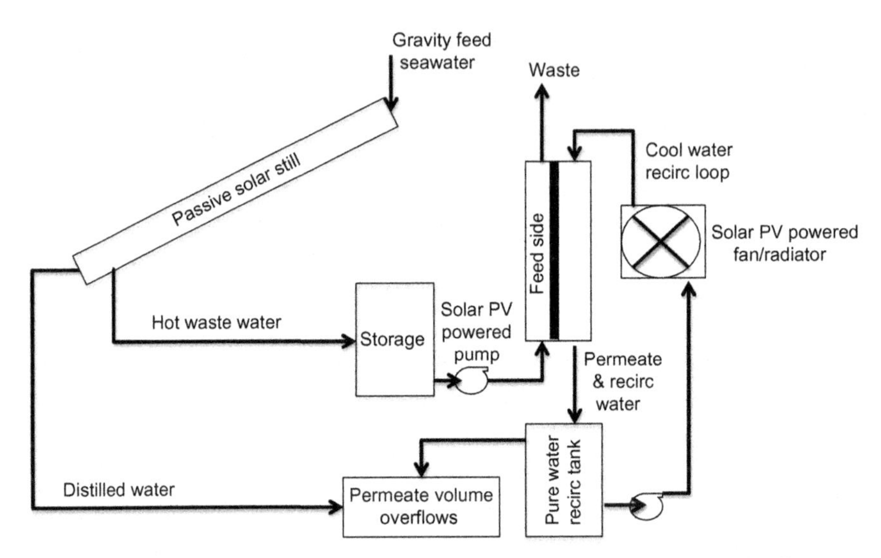

Figure 7.12 Pervaporation membrane system and passive solar distiller. GN enhanced membranes would increase flux rates. *GN*, graphitic nanofiber.

GN can also help increase the strength of polymer desalination membranes. Desalination by pervaporation is a viable process, but a big hurdle lies in maintaining the strength of the membranes with thinner layers, necessary for performance enhancement. Enhanced strength can allow thinner membranes, which would reduce the distance of the gas diffusion layer from the bulk feed flow, consequently keeping the temperature differential between the feed side and the permeate side high.

7.4.2 Heavy metal and ionic contaminant removal

Wastewater from many industries contains toxic heavy metals such as copper, cadmium, lead, chromate, mercury, etc. In addition, some groundwater is contaminated by toxic ionic contaminants such as arsenic, selenium, fluoride, radium, etc. Systems to remove these contaminants constitute a major part of the industrial wastewater treatment industry.

Having spent more than two decades in wastewater treatment, I have tried my best to follow the development of heavy metal removal methodologies in literature. GO and functionalized GO have been investigated for heavy metal removal by many [435–466]. The literature does not give me any cost-to-benefit ratios which I would consider any better than the incumbent technologies, notably ion exchange resins and functionalized adsorbents. Ion exchange resins are polystyrene or polymethacrylic acid polymers mostly crosslinked by divinylbenzene. The largest use for these materials is for deionization and softening of water. These polymers can be functionalized to have a quaternary amine group, tertiary amine group, sulfonic group, carboxylic group, phosphonic group, ethylene diamine group, thio groups, bis-picolyl-amine group, pyridine group, and many more. The last five groups are functional groups typically found in what are known as chelating resins. They have a strong affinity for transition and heavy metals. From the publications on nanomaterials with regards to heavy metal removal, whenever capacities were given in proper units, I converted them to milli equivalents per liter for an apple to apple comparison. Table 7.9 lists the results from work done with GO and functionalized GO for heavy metal removal. The last reference and capacities are for materials already available commercially.

From a cost perspective, there does not seem to be a compelling reason to develop GO and functionalized GO membranes or media for heavy metal removal from aqueous streams except when exceptionally fast kinetics are desired. Otherwise, ion exchange is the dominant technology for that application. It already delivers much better performance in the field than most of the reported results. The external surface area of the GN does provide an advantage, but the current costs are not favorable. Pellenbarg et al. [467] have quantified the functional groups on oxidized herring bone GN using the fluorescence labeling technique. They report 0.40 μmole/g hydroxyl, 0.45 μmole/g carboxyl and 0.90 μmole/g of aldehydes and carbonyl groups.

7.4.3 Arsenic removal

Arsenic appears as a natural toxin in many soils around the world. Bangladesh has been in the news about this problem, but many areas of the world have groundwater contaminated with Arsenic. The developed countries install treatment plants for their communities or consumers can afford replaceable cartridges, such as some of the states in the western United States. The issue has always been the operating cost of the system, and many times the capital cost. Capital costs are often covered by NGOs for underprivileged regions, but the local population is unable to maintain the systems mainly due to costs. Of the literature I have studied about the use of GO and functionalized GO for removal of Arsenic [435,468–476], I could not see any commercial viability. The capacities given [477] (35.83 mg/g and 29.04 mg/g of As (III) and As (V),

Table 7.9 Heavy metal removal data from functionalized GO

Reference	Metal	Meq	Functional group and weight
[456]	Co (II)	0.72	Per gram—GO−COOH
[457]	Pb (II)	0.85	Per gram—GO−COOH
[457]	Cu (II)	0.68	
[457]	Cd (II)	0.70	
[457]	Ni (II)	0.78	
[457]	Pb (II)	0.95	Per gram—1-(3-aminopropyl)pyrrole
[457]	Cu (II)	0.85	Highest values taken
[457]	Cd (II)	0.83	
[457]	Ni (II)	0.84	
[116]	Cu (II)	0.38	Per gram—GO−triethylenetetramine−methacrylate
[458]	Cu (II)	1.34	Per gram—GO-L-tryptophane
[460]	Cr (VI)	0.96	Per gram—PPy/*a*-cyclodextrin/GO
[458]	Pb (II)	4.46	Per gram—GO−COOH
[461]	Cd (II)	0.19	Per gram—GO−carboxymethyl cellulose
	Co (II)	0.54	
	Cu (II)	0.60	
	Ni (II)	0.55	
[462]	Cu (II)	0.88	Per gram—sodium alginate−GO
	Pb (II)	0.32	
[463]	Cu (II)	0.68 ca.	Polydopamine−polyethylene amine−GO
	Cd (II)	0.47 ca.	
	Pb (II)	0.48 ca.	
	Hg	0.55 ca.	
[464]	Pb (II)	1.23 ca.	Hydrated manganese oxide−GO
[465]	Cu (II)	1.3 ca.	Fe_3O_4
[466]	Cu (II)	1.34	Commercial ion exchange resins Functionalized with imminodiacetic groups
	Cd (II)	1.10	
	Ni (II)	1.34	
	Zn (II)	0.95	
	Hg (II)	0.71	Commercial ion exchange resins Functionalized with thiouronium groups
	Cr (VI)	1.00	Commercial ion exchange resins Functionalized with quaternary amines

respectively), [478] (77.5 mg/g As (III) and 45.7 mg/g (As V) using akaganeite chelate) do not seem to offer much of a cost-to-benefit ratio compared to other media we use in the water treatment industry. Iron oxide coated sand [479,480] and filter fibers coated with iron oxide offer the same capacities.

7.4.4 Adsorption

Activated carbons are the most widely used sorbents in industry for dechlorination, deozonation, and organic contaminants removal from water streams. Their capacity for adsorption is directly related to the porous structures and the presence of surface chemical groups. The most common groups are carboxyl, lactonic, carbonyl, and phenolic functionalities. While activated carbons are effective for the adsorption of organic molecules, these materials can indiscriminately adsorb not only the contaminant but also water molecules, and consequently, the available volume for the undesirable organic molecules is greatly reduced. The accessible surface area in activated carbon has been discussed in the catalysis section. The common shortcoming of activated carbon is its very limited ability to remove nonpolar compounds, or charged compounds that behave like ions. In wastewater treatment, removing textile dyes and pharmaceutical compounds without subjecting the water to flocculating chemical processes presents a challenge. The wastewater from textile mills is not always toxic to humans, but the color blocks sunlight and affects marine life. The capacity of activated carbon is low for these contaminants, and it cannot be regenerated efficiently. Adsorbents that can effectively and affordably address the mixed nature of these wastewaters are really desirable right now. Most biomass wastes can be used to adsorb dyes, due to their partial ionic nature, but these materials are not amenable to an industrial process design. Gregorio [481] provides an extensive list of such materials. GO enhanced adsorbents reported in literature sound promising for an economical solution for this growing problem and are worthy of further investigation toward commercialization. One material that is abundantly available as waste is chitosan. We have discussed chitosan and the possibilities of GN chitosan hybrids. Chitosan is the deacetylated derivative of chitin, which can be easily be extracted from crustacean shells. Chitosan can be used as an adsorbent to remove ionic dyes due to the simultaneous presence of amino and hydroxyl groups. GN can improve the performance of chitosan as an adsorbent if it is used as a cross linking agent. Travlou et al. [482] synthesized graphite oxide–chitosan composite (and the authors emphasize "graphite" not graphene) (GO–Ch), which consisted of crosslinked chitosan (Ch) and GO. Reactive Black 5 dye was used to study the adsorption experiments. 277 mg/g uptake of Reactive Black dye, was reported, which is several times that of activated carbon (59 mg/g). Gonzalez et al. [483] also report a chitin–GO composite for the removal of Remazol Black and Neutral

Red dyes, with very encouraging results. Jiao et al. [484] report magnetic Fe_3O_4 deposited on GO for Rhodamine removal. Given the design limitations of using GN and nano size materials, an adsorbent that can be magnetically separated and regenerated is very attractive to engineers. Polyaniline-coated GO has been tried successfully to remove cationic and anionic dyes from water [485].

Aside from dyes, other difficult to remove compounds can be removed from wastewater by GN. Oxidized GN have been shown to remove naphthalene, 2-naphthol, and 1-naphthylamine and one pharmaceutical compound (tylosin) by Ji et al. [486]. If magnetized by deposition (not bonding) of Fe_3O_4 as discussed earlier [472], this could bring very large volume applications for GN in a field that is growing. Unlike activated carbon, GN adsorb organic molecules preferably over water molecules. Park et al. [487] have studied the selective removal of organic molecules, especially the difficult to remove low molecular weight alcohols from aqueous solutions using herring bone and platelet-type GN. A significant expansion in the lattice spacing of the GN was observed following adsorption of butanol, which suggests that the organic molecule "slides" between two graphene layers, causing the structure to undergo an expansion. A significant observation to note is that the more disordered "herringbone" structures have a relatively short stack height. The various acid treatments did not affect the *d* spacing of the GN but did tend to make them shorter (Table 7.10).

Adsorption of ethanol on active carbon and GN treated with 1 M HCl was studied by the same authors. The results showed that the ability of the GN to adsorb the alcohol far exceeded that of activated carbon. The uptake of ethanol on activated carbon was only 6%, whereas 38% was achieved on the GN, even though the activated carbon had much higher reported surface area. It is apparent that adsorption of the organic from an aqueous solution is nonselective on activated carbon: An indication of the low affinity toward the alcohols. While in the case of GN, which consists entirely of graphitic platelets, there is proof that narrow molecules "slide" between the graphene layers where they interact with the basal plane regions of the material. Water molecules have negligible affinity for the basal plane of GN and so do not have access to the inner structure. The GN removed 75% of the butanol from the solution, whereas the activated carbon adsorbed only about 28%. Apparently, GN have two main regions that result in an array of hydrophobic nanopores with slightly polar edges, resembling a micellar structure that has a high affinity for nonpolar substances while suspended in, and in intimate

Table 7.10 Characteristics of GN used for ethanol removal

Type of fiber	Treatment	Mean *d* spacing (Å)	Platelet stack height (Å)
GN	Pristine	3.35	175
GN	HNO_3 treat	3.36	128
GN	HCl treat	3.38	138
GN	H_2SO_4 treat	3.40	156
GN	HCl treat/ EtOH adsorption	3.38	140
GN	HCl treat/ butanol adsorption	3.44	108
Herring bone GN	Pristine	3.47	54
Herring bone GN	HNO_3 treat	3.46	47
Herring bone GN	HCl treat	3.47	46
Herring bone GN	H_2SO_4 treat	3.49	56

Source: Data reproduced from N. Pugazhenthiran, S. Sen Gupta, A. Prabhath, M. Manikandan, J.R. Swathy, V.K. Raman, et al., Cellulose derived graphenic fibers for capacitive desalination of brackish water. ACS Appl. Mater. Interf., 7(36) (2015) 20156–20163.

contact with aqueous media. In this case, derivation of the comparable costs would not be straightforward. On the surface, a 6× capacity at a 25× price difference does not make sense. But in wastewater treatment, the ease of regeneration, number of cycles before disposal and disposal costs all play a critical role in the analysis.

Taking Park's data as an example:

- High quality wood powdered activated carbon (PAC) may cost $3.0 per kg.
- GN cost $51 per kg.
- Activated carbon adsorption of EtOH ~ 6% by weight.
- GN adsorption of EtOH ~ 38% by weight.
- Activated carbon cycles of use ~ 5
- GN cycles of use ~ 30
- Hazardous waste disposal ~ $200/MT
  - ~100 g EtOH adsorbed by 1.67 kg-PAC
  - ~100 g EtOH adsorbed by 0.26 kg-GN
- 500 g-EtOH adsorbed in 5 cycles PAC.
- PAC cost ~ ($3/kg × 1.67 kg) = $ 5 per 500 g-EtOH.
- 3,000 g-EtOH adsorbed in 30 cycles.
- GN cost per 500 g-EtOH = 0.26 kg-GN/3,000 g-EtOH × 500 g-EtOH × $51/kg-GN = $ 2.21 per 500 g-EtOH

Therefore GN costs less than half of PAC in removing EtOH from aqueous solutions, if the data from Park et al is reproducible.

7.4.5 Capacitive deionization

7.4.5.1 Capacitive deionization general

Capacitive deionization (CDI) desalinates water with an electrical potential difference over two high surface area electrodes. Due to the electric potential, negatively charged ions (anions), are transported to the surface of the positively polarized electrode, the anode. Likewise, positively charged ions (cations) are transported to the negatively polarized electrode, the cathode. CDI technology is best suited for desalinating brackish water or partially deionizing municipal water with high levels of dissolved solids ($\sim 1000+$) as a pretreatment for reverse osmosis systems. Compared to reverse osmosis and distillation, CDI is more energy-efficient and can be configured for very high recovery rates. It's productivity (volume of water purified between cycles) is simply a function of the Faradaic capacity of the electrodes.

The operation of a conventional CDI system cycles through two phases: An adsorption phase where water is desalinated and a desorption phase where the electrodes are regenerated. During the adsorption phase, a potential difference over two electrodes is applied and ions are adsorbed from the water. The ions are transported through the interparticle pores of the porous carbon electrode to the intraparticle pores, where the ions are electrosorbed in the so-called electrical double layers (EDLs). Briefly, in EDLs, there are basically three distinct layers.

1. A porous conductive matrix with the electrical surface charge.
2. A diffusion zone, in which the counter ions balance the electrical charge of the electrode matrix.
3. A dielectric layer located between the electrode matrix and the diffusion zone. Referred to as the Stern zone, it separates the two charged zones, but does not carry any charge itself.

Any amount of charge (potential) will always attempt to be neutralized by the same amount of counter-charge. During the adsorption cycle, a charge is created on the surface area of the conductive electrode. This charge must be balanced by an opposite charge. This is achieved by the charged ions in water.

When the electrodes are saturated with the relative ions, the adsorbed ions are released for regeneration of the electrodes. The potential difference between electrodes is reversed or reduced to zero (grounded). In this way, ions leave the electrode pores and can be flushed out of the

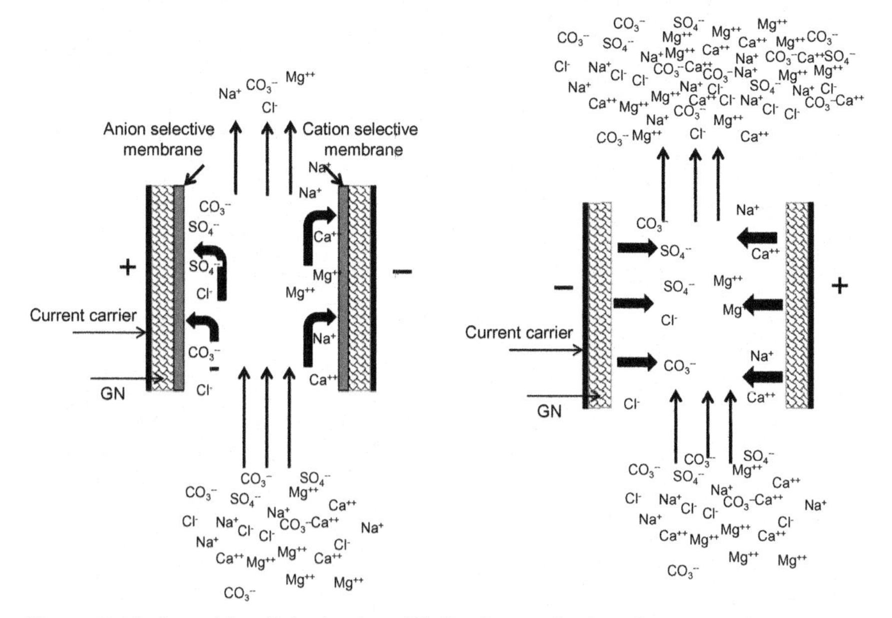

Figure 7.13 Capacitive Deionization. (A) Service cycle. Ions in aqueous stream are attracted to the GN electrode through an ion selective membrane for the respective electrodes, resulting in desalting of the stream. (B) Regeneration cycle. When potential is reversed or grounded, ions are forced away from electrodes and exit as a concentrated stream, resulting in ion free electrodes.

CDI cell resulting in an effluent stream with a high salt concentration, normally called the brine stream or concentrate. Part of the energy input required during the adsorption phase can be recovered during this desorption step by charging a battery instead of grounding the electrodes. A discussion on that subject is beyond the scope of this book.

7.4.5.2 Membrane capacitive deionization (MCDI)

Due to the EDL phenomenon, co-ions in one diffusion zone will tend to travel to the zone next to the opposite electrode. If both electrodes are covered with ion selective membranes, co-ions cannot travel toward the opposite electrode, and more counter ions are needed to balance the charge. This configuration is called membrane capacitive deionization (MCDI), as shown in Fig. 7.13.

7.4.5.3 Modes of operation

a. Constant voltage. The voltage is kept constant during the adsorption cycle. In this mode, the effluent salt concentration decreases, but after a while, it goes up again. As the diffusion zone starts accumulating counterions, the potential at the electrode decreases, and new counter ion transport rate goes down, leaking the salt ions in the effluent.

b. Constant current. As the ionic charge transport to the electrodes always equals the magnitude of the applied current, applying a constant current keeps the ion removal stable. This mode is used for (MCDI).

7.4.5.4 Cell formats

a. Plate & Frame type: The electrodes are placed in a stack and the water flows through the spacer between them.

b. Flow through type: The current carriers are alternating stacked Grafoil sheets coated with the high surface area electrode material, and then with the ion selective membranes. The feed water flows from the outside boundary of a MCDI, travels through the spacers between each electrode pair and exits through a center aperture.

c. Flowing electrode type: Similar to flow through type MCDI, but instead of having solid electrodes, a carbon suspension (slurry) flows between the membranes and the current collector. A potential difference is applied between both channels of the flowing carbon slurries. As the carbon slurries flow, the electrodes do not saturate, and therefore, this cell design can be used for the desalination of water with high salt concentrations as well. A regeneration cycle is not necessary. The slurries are mixed together and separated from the salt water.

7.4.5.5 Electrode materials—the role of GN

Currently, mesoporous-activated carbon electrodes are used as the electrode materials. The critical quality of the electrode material is the specific surface area. The accessible surface area, that can be charged and allows access to the counterions, must be large and relatively unobstructed. The structural difference (with regards to accessible surface area) between activated carbon and GN has been discussed before. GN are a natural for this application. Pugazhenthiran et al. [488] used cellulose derived graphitic fibers with calculated BET surface area of 598 M^2/g. The electrode showed an electroadsorption capacity of 13.1 mg/g of the electrode at a cell potential of 1.2 V. The test water was 500 mg/L of sodium chloride, which is not the typical application range of CDI, and it is not clear what the percentage removal of NaCl was (inlet−outlet). Li et al. [489] used a reduced GO and activated carbon composite and report almost double the electrosorption capacity of activated carbon alone (at 2 V ~0.85 mg/g vs 0.4 mg/g, respectively). The reason given is the formation of intricate bridging between the pores of the activated carbon with the reduced GO. The conductivity of the feed water is given at 50 μS/cm. Two factors can be interpreted from the above. The higher the surface area, the higher the specific capacitance, and the higher the feed water salt level, the higher the capacity. Higher concentrations would give higher specific electrosorption capacity at the expense of lower percentage ion removal during the adsorption cycle, because there will undoubtedly be slippage. In system design, an alternating cycle lead-lag system can achieve a higher overall efficiency than a single stage once through system. The data [488,489] surely gives proof that GN will enhance performance compared to activated carbon. But for practical evaluation, it would be nice to know all the parameters.

My team's experience is stated in more general terms. This is a technology our team has spent considerable time and resources on. Generally speaking, herring bone type as produced GN electrodes in a single stage once through MCDI system gave 30% better performance in once through salt removal efficiency (inlet TDS−outlet TDS) and 85% higher specific electrosorption capacity than activated carbon electrodes. The resulting water recovery was ~83−88% during a beta test for several months. City water from the greater Los Angeles area equal to 800−1120 μS was used. Average outlet conductivity was <100 μS for equivalent weights of the GN and activated carbon. Platelet type and ribbon-type GN performed only marginally better than activated carbon.

7.5 GAS PURIFICATION

7.5.1 As adsorbents

GN show promise in significantly improving the performance in remediation of waste vapors as superior adsorbents. Many organic vapors are removed by adsorption in industry. Manufacturers of specialty chemicals and pharmaceuticals generate effluent streams that contain trace amounts of aromatic and chlorinated hydrocarbons. In addition, contaminated groundwater from decades of unlined or leaking storage tanks is being treated by "pump and treat" methods all over the United States. Associated with these contaminated groundwater streams are soils, which are contaminated with the same organic constituents, such as tri-chloro-ethylene (TCE) per-chloro-ethylene, gasoline, diesel, jet fuel, etc. If contaminant volumetric concentration is high, the soils are typically treated with either vacuum systems or volumetric replacement of vapor, followed by catalytic combustion or removal/recovery by adsorbents. If the contaminant levels are low, adsorbents are used to concentrate the contaminants. Careful handling, recovery, or disposal of these toxic organics is one of the major environmental issues that confront such industries. Methods for the elimination of such contaminants from gaseous and liquid effluent streams are normally based on fixed-bed adsorption on carbonaceous materials.

Efficient desorption from the media and recovery of the chemical is sometimes desirable, if the product can be reused with minimal reprocessing. The regeneration process for volatile organic compounds is carried out with steam. The high temperature steam volatalizes the organic contaminant and carries it to a condenser where the contaminant is recovered in a much higher concentration. During steam regeneration, activated carbon absorbs water and some hydrophobicity is lost in every regeneration cycle. The high external specific surface area of as produced GN would offer an advantage in this case and facilitate a very high percentage recovery of the adsorbed material. The inherent hydrophobicity makes the GN perform better during the adsorption and desorption cycles, as the surface of the as produced GN is hydrophobic.

7.5.2 For catalytic remediation

I have discussed the catalytic qualities of the GN without any metal catalyst. The π electrons and hydrophobicity make the GN ideal candidates for catalytic degradation of organic and inorganic contaminants into less toxic products at lower temperatures. Unlike the front-end production processes, remediation is a cost center for industry, and any cost reduction is valuable. An effective catalyst is one on which the contaminant is initially strongly adsorbed. Then a reaction with atomic species generated by the interaction of metallic components with the aqueous environment takes place.

An important toxic contaminant that is destroyed in this way is H_2S. It originates predominantly from the extraction of natural resources such as oil, gas (sour gas), coal, and geothermal vapor. H_2S is also present in biogas in various concentrations depending on the source of the organic material being digested. Anaerobic digestion of manure tends to have high levels of sulfur compared to the biogas from landfills. Low levels of H_2S in the gas stream are typically treated by adsorbents which use some form of iron to precipitate the insoluble FeS. These adsorbents typically exhibit a low weight/weight capacity (25–30%), and end up creating a significant volume of waste.

For large flows and higher mass fractions of H_2S, it is desirable to oxidize it to elemental sulfur and return it to the industrial loop. The multistage Claus process is used for the production of sulfur from an H_2S-containing gas.

$$2H_2S + O_2 = 2S + H_2O$$

In the Claus process, H_2S-bearing gas enters at low pressure (<10 psig) and low temperature (120°F). The gas is mixed with a compressed air stream which then flows to a burner and is fired into a reaction furnace. This free-flame Modified Claus reaction can convert approximately ½ to ¾ of the sulfur gases to sulfur vapor. The exit gases can be as hot as 1400°C. The gases are cooled in a steam generating heat exchanger first, and further cooled to approximately 165°C, which condenses the sulfur. The resultant liquid sulfur is removed and kept molten in a storage tank. Multistage catalytic reactors with precious metal catalysts are deployed as a second stage, which remove more sulfur, and a tail gas down to 1–2% sulfur can be produced. The tail gas is sent to polishing adsorbents, catalytic destruction, water scrubbers, etc.

GN or other carbonaceous catalyst supports cannot be used at these temperatures for the multistage catalytic sulfur recovery in the presence of oxygen. However, of interest to us is the tail gas treatment of these gases, as well as the desulfurization of biogas and landfill gas. The tail gases of the Claus process may contain 0.4–1.5 vol% of H_2S and up to 40% water vapor. I propose GN can be used to perform in situ Redox reactions. Fe^{+++} ion can be immobilized in a chelating ligand by using techniques discussed earlier for heavy metal removal. The H_2S gas can be contacted with GN loaded on graphitic fiber felt and functionalized with a chelating agent. Iron can be chelated as Fe^{+++} after sequential treatments. H_2S can be passed through the vessel with high humidity. The H_2S is oxidized to elemental sulfur and hydrogen ions (protons).

$$H_2S + 2Fe^{+++} = S^0 + 2Fe^{++} + 2H^+$$

For the regeneration, oxygen with high moisture content, or humid air, can be passed through the vessel:

$$2Fe^{++} + 2H^+ + \frac{1}{2}O_2 = 2\,Fe^{+++} + H_2O.$$

The process could solve issues associated with the same concept, which is typically done in the liquid phase. Gendel et al. [490] report a low pH (1.0) liquid Redox process in which the regeneration is achieved by electrolytic means. Nagl [491] explains a commercial liquid redox process offered by Merichem. The process uses a liquid containing Fe^{+++} ions chelated with EDTA or other suitable chelating agents. The regeneration is performed by air or oxygen, the same mechanism explained above. The process is reported to have very short life for the chelating agent [492].

Other catalytic carbon materials are manufactured with oxidation to create defects and reactive edges which act as catalytic sites resembling those of GN. These high activity sites are used to oxidize H_2S to elemental sulfur in the presence of moisture. Calgon carbon markets a catalytic carbon with the brand name Centaur® that is used for landfill gas and biogas purification. Verbal communication in the past with Calgon informed me the upper limit of H_2S in the feed stream was 100 ppm (vol %). There are some theoretical models published for use of GO for this function [493].

7.5.3 For separations

GN can increase adsorption performance of carbonaceous adsorbents in PSA (Pressure Swing Adsorption). PSA systems are used extensively in industry for gas separations. In a PSA, the feed stream is put into contact with a packed fixed bed adsorbent at an elevated pressure. Activated carbon is used preferentially in PSA systems due to its large surface area. Gas sorption, in activated carbon, is via physisorption on the surface and is affected by electrostatic and dispersion forces. Electrostatic forces can be controlled by a charge on the adsorbent, while dispersion forces can be controlled by chemical substitution. The strength of the interaction is determined by the surface characteristics of the activated carbon and the properties of the gas to be adsorbed, such as size, dipole moment, polarizability, magnetic response, dew point, etc. Gadipelli et al. [494] list Table 7.11 from work done by Li et al. [495] which summarizes the adsorption-related physical parameters of many gas or vapor adsorbates. For instance, an adsorbent with a high specific surface area is a good candidate for adsorption of a molecule with high polarizability but no polarity. Adsorbents with highly polarized surfaces are good for adsorbate molecules with a high dipole moment. The adsorbents with high electric

Table 7.11 Some of the physical parameters of selected gaseous adsorbates

Adsorbate	Normal BP (K)	T_C (K)	V_C (cm^3/mol)	P_C (bar)	Kinetic diameter (nm)	Polarizability $\times 10^{25}$ (cm^3)	Dipole moment $\times 10^{18}$ (esu cm)	Quadrupole moment $\times 10^{26}$ (esu cm^2)
He	4.30	5.19	57.30	2.27	0.2551	2.04956	0	0
H_2	20.27	32.98	64.20	12.93	0.2827–0.289	8.042	0	0.662
N_2	77.35	126.20	90.10	33.98	0.364–0.380	17.403	0	1.52
O_2	90.17	154.58	73.37	50.43	0.3467	15.812	0	0.39
CO	81.66	132.85	93.10	34.94	0.3690	19.5	0.1098	2.50
CO_2	216.55	304.12	94.07	73.74	0.33	29.11	0	4.30
NO_2	302.22	431.01	–	101.00	–	30.2	0.316	–
SO_2	263.13	430.80	122.00	78.84	0.4112	37.2–42.8	1.633	–
H_2S	212.84	373.40	98.00	89.63	0.3623	37.82–39.5	0.9783	–
NH_3	239.82	405.40	72.47	113.53	0.2900	21.0–28.1	1.4718	–
H_2O	373.15	647.14	55.95	220.64	0.2641	14.5	1.8546	–
CH_4	111.66	190.56	98.60	45.99	0.3758	25.93	0	0

Source: From R.T. Yang, Adsorbents: Fundamentals and Applications, John Wiley & Sons, Hoboken, NJ, 2003.

field gradient surfaces are found to be ideal for the high quadrupole moment adsorbate molecules [496].

At the specific pressure, the less adsorbed component moves through the column faster and the strongly adsorbed component(s) will see a retardation effect due to adsorption and slower movement from the attractive forces. Separation is never discreet, and the breakthrough of the strongly adsorbed component will occur slowly at first, then a rapid rise will be experienced. At a predetermined leakage level, the bed is taken offline and the flow diverted to another column. The pressure on the bed is released and the direction of the effluent is changed to a different tank. The column is purged with the desired component and pressurized again. The entire cycle can be only few minutes long. For the reasons described in the adsorbent section, GN are well suited for this application [494] because specific surface area of the adsorbent is critical to enhance gas adsorption, though the GN would have to be coated on a honeycomb to avoid a large pressure drop. Other methods such as VPSA (vacuum pressure swing adsorption) exist, but are not used commercially on a similar scale.

Thermal swing adsorption is similar to PSA except desorption is achieved by heating the adsorbent. GN have also been shown to enhance the performance of pressurized gas adsorption, such as the adsorption of CO_2 [497–499] with a desorption cycle consisting of heating of the adsorbent.

The high electrical conductivity of GN makes them good candidates for a low energy consuming separation process called electric swing adsorption (ESA). ESA is a gas separation process similar to the pressure and thermal swing processes. ESA achieves desorption via an external electric perturbation [500].

7.5.4 Summary

The key issue discussed was the ability of oxidized and as-produced GN to perform better than carbon in adsorption applications. With external surface area and high-energy π-π bonding capability, GN can outperform activated carbon adsorbents in kinetics and capacity.

The edge zones act as catalytic sites for oxidation at lower temperatures for environmental applications. A new redox method of H_2S oxidation to elemental sulfur using immobilized and regenerable oxidation agent was briefly mentioned. H_2S can be oxidized and S removed as a solid, while the oxidant can be regenerated using an oxygen source such as air. GN seemed to make this process viable.

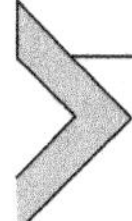

7.6 HEAT TRANSFER & LUBRICATION

7.6.1 Heat transfer

GN are essentially nanostructured crystalline graphite, which is a well-known and extensively used additive for electrical conductivity. Graphite's performance as a heat transfer aid is intuitively obvious. Sufficient heat transfer is critical for the productivity and life of the equipment used for heat exchange, not only for the instantaneous benefit at the heat transfer surfaces, but also for the eventual size of the equipment, if the cooling cycle is composed of recirculating fluids. Better heat transfer translates to lower energy costs in recirculating fluids because lower volumes and flow rates would be required. The heat transfer efficiency of conventional fluids like deionized (DI) water, EG, transformer oil, etc. is enhanced by the addition of GN. These and other liquid heat transfer media have a lower thermal conductivity compared to solid heat transfer media. Addition of thermally conducting solids to these fluids improves their conductivity, provided the size of the solids does not present a viscosity or deposition problem, and the solids do not react adversely with the fluid. Addition of 1% by volume of CNTs into poly (α-olefin) oil by Choi et al. [501] increased the thermal conductivity by a ratio of 2.5:1 to the original base fluid. Theory had predicted only a 10% improvement over the base oil conductivity. They attributed the additional thermal conductivity enhancement, to the way the CNT organized at the liquid solid interface. We have seen this property in our discussion of conductive plastics. Nanoparticles of metals and their oxides have been added to heat transfer fluids but not relevant to our discussion.

There is more to the story, though. Enhancements with GN are not linear, and engineering input will have to play a big part in the implementation of this application. Beyond a certain solids concentration, additional solids in the fluids would increase the viscosity. Viscosity increases would result in increased pressure drops [502,503], which, in turn, would throw the cost versus benefit ratios off. In addition, the effects of axial distance on the heat transfer efficiency must be taken into account. There is also conflicting research on increase [504] and decrease [505] of thermal conductivity with increasing axial distance of a flowing fluid from the heat transfer interface. Either way, it will have to be one of the important parameters for which empirical data would need to be collected and evaluated before commercial aspirations. So, is

Table 7.12 Summary of data provided calculated costs and benefits

Reference	Base material	Base material price (US $/MT)	Type nano-material	Dosage	Cost of enhanced product (US $)	Cost increase (%)	Reported improvement (%)
[506]	Propyl glycol	1300	CNT	5 vol% 1.2 wt %	2400/MT	85	62.3
[507]	DI water	3.00	CNT	0.25% by weight	221/MT	10,000	150
[501]	Mineral oil	5,000	CNT	1 vol% 0.2 wt%	5190/MT	3.8	160

it viable to propose enhancement of heat transfer fluids by carbonaceous nanomaterials? Let us have a look.

The loading percentages discussed by some authors is given in Table 7.12. The costs for current materials and cost of materials used are taken from Table 5.1, Percentage improvement in thermal conductivity is presented to have a cursory view of the cost-to-benefit ratio.

As common sense would dictate, the higher the value of the base fluid, the more attractive the improvement in thermal conductivity looks. The cost data can be cross-referenced with the viscosity and heat transfer characteristics in axial flow (discussed above) and a better decision can be arrived at for the mineral oil application.

7.6.2 Lubricants

Lubrication is a natural application for graphitic materials. Graphite itself is one of the best solid lubricants around. Given that the lubrication provided by graphite is due to the sliding effect between parallel planes, it is amusing to sometimes read about "graphene" enhanced lubricants. Hopefully they are not referring to single layer graphene. The improvement in lubrication qualities as well as improving the life of the lubricant by providing better thermal conductivity is a very attractive prospect for the industry. For evaluating performance of lubricating fluids, the typical parameters used in the tribological studies and standards are friction coefficient (FC) and wear scar diameter (WSD). The tribological property improvements with addition of nanographitic materials, including ball-milled graphite has been investigated. Zhang et al. [508] report a vegetable oil modified with "graphene" displaying a 17% improvement in FC and 14% improvement in WSD by a very low dose of 0.02–0.06% by weight of hydrazine reduced GO. Eswaraiah et al. [509] achieved a 33% improvement in WSD and 80% reduction in the FC with solar concentrated heat exfoliated GO. Similar results (33% improvement in WSD) were achieved by Huang et al. [510], but by using 0.01 wt% nanographite. Interesting to note that they simply ball-milled flake graphite down to 10–20 nM to achieve these results. Senatore et al. [511] used GO instead of graphite flakes, and their results did not compare favorably with the earlier results cited. They studied the FC and WSD improvements in three regimes (layers). i) The boundary layer (closest to the moving parts), ii) Mixed layer (transition from metal/metal to no contact), and the iii) Elastohydrodynamic layer, where viscosity is highest. The FC varied from 20% in the boundary layer to 8% in the elastohydrodynamic layer. WSD improved by 12%, 27%, and 30% in the boundary, mixed, and elastohydrodynamic layers respectively.

Let us take a look again at the viability of these research efforts with the industrial world view, in the near future. The loading percentages discussed by some authors is given in Table 7.13. I have calculated the costs of current materials, what it would cost to enhance their performance by the stated percentages using the values from Table 5.1.

The results, taken at face value look promising, as expected from the discussion, but my contention is that the volume of GN required for this application is not something that can be accommodated in the near future. Millions of tons of material are produced every year. A discouraging sign is the negligible associated cost increase using ball milled graphite has not been enough of a driver for industry to adopt this concept. This is the reality of commoditized industries adopting new concepts. The gestation period is multiple times more than the length of time any non revenue generating startup can survive.

7.6.2.1 Minimum Quantity Lubricants (MQL)

On the optimistic side, lubrication has new frontiers that can be exploited. I discussed earlier that a good market entry point for GN and other nanomaterials is into markets that have not yet reached the inflection point on the "hockey stick" curve. In this category, one such application that can be targeted is the minimum quantity lubricant market sector. In the pursuit of more environmentally-friendly manufacturing, alternatives to flooded cooling are being sought in the machining industry. Dry machining is one such option and has helped some machining operations eliminate cutting fluids completely because of the rapid progress in tool coatings [512]. However, by and large, cutting fluids are still required due to the high friction and adhesion tendency between

Table 7.13 Summary of data provided in literature, calculated costs from data, and quantified improvement in characteristics using values for cost of manufacturing from Table 5.1

Reference	Base material	Base material price	Additive	Dosage (wt%)	Increased cost	Cost increase (%)	Performance improvement
[508]	Poly Alpha Olefins	$ 1,500/MT	Reduced Graphene Oxide	0.045	$1,660/MT	11	FC = 17% WSD = 14%
[509]	Group II engine oil (assumed)	$ 650/MT	Conc. Solar Reduced Graphene Oxide	0.003	$656/MT	9	FC = 80% WSD 33%
[510]	Group II engine oil (assumed)	$ 650/MT	Ball Milled Graphite	0.01	$ 651MT	Insignificant	WSD = 33%
[511]	Group I base oil (assumed)	$ 250/MT	Reduced Graphene Oxide	0.1	$ 840/MT	>350	Max: FC = 20% WSD = 30%

work and tool materials, which increases the difficulty in chip and heat removal [513,514]. An optimal solution, wherever suitable, is minimum quantity lubrication (MQL) as a viable solution to overcome the disadvantages of both the dry and flood cooled machining processes. MQL is known by several names currently, as is the case with any new approaches (sound familiar?). It can be called "near-dry machining" or "NDM," "micro-lubrication," "micro-dosing," and sometimes "mist coolant." In a nutshell, MQL makes use of a lubricant, not a coolant, and does so in "minimum quantities." In flood cooling, the coolant uses the fluid to provide cooling. MQL instead coats the interface with a thin film of lubricant and prevents heat buildup through friction reduction. The high lubricity of a good MQL helps transfer a large fraction of the frictional heat to the chips as they are expelled from the machine. This action keeps the cutting tool much cooler and reduces tool wear. MQL lubricants are mostly consumed in the cutting process. The friction and heat in the interface vaporizes the small amount of lubricant and leaves cutting tools, parts, equipment, and floors dry and clean. In their work, Park et al. [512] report the addition of GN with a diameter of 1 μM producing a lubricant which was stable for more than 1 year with vegetable oil as a base. Vegetable oil advances the green impact further by originating as a renewable source. This application is expected to grow, and provide an opportunity for GN to get integrated with the opportunity.

7.7 DRILLING FLUIDS (DF)

DFs are a critical part of the oil production process. They are also called drilling muds. This is a complex material to discuss in terms of applications. A brief introduction would help understand the references cited, and my comments to the same. There are three major kinds of DFs:

Water-based—Saltwater

Water-based—Freshwater

Oil-based—Including synthetic base, such as GTL-derived fluids.

Probably, the highest priority in terms of selecting a DF (not including environmental constraints) is COST. If there is no significant cost-to-benefit ratio, carbonaceous nanomaterials are dead in the water.

Water-based fluids are used in probably 80% or more sites in the world. They are the most widely used systems because they are less expensive than oil-based fluids (OBFs) or synthetic fluids. OBFs technically can be called invert-emulsion systems. Invert emulsion systems have an oil base fluid as the continuous phase, and high salinity water as the internal phase. Such fluids are very expensive and are only selected when absolutely necessary to preserve shales that might otherwise get damaged due to lower lubricity of the water-based fluids. Both water-based systems and invert-emulsion systems can be made to function at high temperatures.

Saltwater DF often are used for drilling into salt formations. The high salinity inhibits hydrates near deep-sea wellheads that affect critical processes. Some of the major ingredients comprising the salts are $CaCl_2$, $CaBr_2$, $ZnBr_2$, and $KCHO_2$.

DFs carry drill cuttings back to the surface, lubricate the bit, provide cooling to the bit, prevent corrosion, and provide a medium to the electronic data acquisition systems about the formation. Important and relevant to an application discussed in this section, DFs provide support to the walls of the well bore and create a protective cake if the reservoir consists of permeable formations. If there is DF leakage into such permeable, porous formations, the reservoirs can be damaged. To prevent this, "fluid loss additives" are added to the DF. These additives are typically flexible materials such as bentonite, lignite, micron-sized rubber particles, etc. The idea is to block the pores during drilling by the pressure of the DF, and when the pressure of the DF is eliminated, the formation oil pressure pushes these materials out. Kosynkin et al. [515,516] have suggested the use of methylated "graphene" oxide as a fluid loss prevention additive. The tests were conducted as per API procedures with hardened low ash grade 50 having an estimated 2.7-μM pore size. The thickness of the filter cake and the fluid loss as filtrate at 100-psig pressure was measured. They produced better results with a mixture of flake size (ranges) than they did with a single range of flake sizes. The best performance they achieved was with a 3:1 mixture of (esterified) large flake GO (LFGO) to graphite powder derived GO (PGO). The composition was prepared based on carbon content, with LFGO and PGO carbon content at 5.2% and 30%, respectively. At 2 g/L, the filtrate rate was 50% worse than the bentonite formulation, but the cake thickness was >90% lower. At 4 g/L, filtrate volume was 15% less than the commercial bentonite formulation. Cost implications will be given later, but the authors emphasize that thinner cakes help avoid pipe sticking, and the lower level of suspended solids increases the rate of penetration. In addition, the GO addition gives the DF good viscoelastic properties. However, no quantification of these benefits are provided, even in the referenced text, so it is hard to determine the financial and downtime impacts from replacing bentonite formulations with GO formulations. Ismail et al. [517] have published results for MWCNT addition for the same application. Unfortunately, they list units of mass when referring to concentration. There is no way of determining if the weight given is per unit volume of DF or per unit weight. The only field trial reported with "graphene" enhanced DF was from Norazasly et al. [518] who presented a paper at the Int'l Petroleum Technology Conference detailing field comparison results for permeability, lubricity and friction using "graphene-enhanced" DFs vs ester-based DFs. Like so many marketing papers presented at technical conferences, this one seems to have a promotional goal only. But in this instance, the authors were not just reporting an incremental improvement in performance, as is done in most conferences. They were supposedly reporting results from an actual *field trial*. Field trials are not only a major expense, but also a huge risk taken by the oilfield services company (in this case, SCOMI). Their results are a major achievement for the nanomaterials industry, proving all the benefits graphene has been touted for. It is also a tool for the oil companies to cut costs. Not withstanding the marketing intent, more non-confidential data on the formation of the oilfield, source of graphene, type of graphene, etc., would only serve to help both industries. It would behoove the company to launch a major drive and promote the formulation. However, to my disappointment, the data is incoherent and conflicts with other work reported, and is not accompanied by any explanation as to why. I have to assume the authors actually mean GO, not "graphene", since the trials were performed with a water-based mud and r-Go would be hydrophobic. As I mentioned, their data conflicts with data presented in [515] from Rice University and MI-Swaco with regards to the stability of the GO in DFs with high salinity. For permeability results, the retained permeability was only marginally better than ester-based lubes (41% vs 49%), assuming the authors mean % in the final row of their comparison tables. The unit given is in millidarcy (mD) which is a formation characteristic. The addition of plain ball milled graphite flakes to oil-based lubricants has shown remarkable improvements in lubricity. Other materials have been reported as new potential fluid loss additives. Patent application US 2015/0008044 A1 [519] outlines an elastomer added at very low dosage reducing the permeability by 77% compared to the base fluid in a water-based mud, see Table 7.14.

Table 7.14 Cost and benefit data comparison for GO additives in DFs

Reference	[515,516]	[517]	[519]
Additive	**Esterified GO from flake graphite**	**MWCNT**	**Elastomer**
Dosage (kg/MT)	0.36% GO (523 provides data on a carbon content basis. GO has a weight gain of 80% from noncarbon groups as a result of oxidation) 1% MeOH for esterification	0.120 kg/T	12% (10.1 lb/gal. Mud)
Assumed specific gravity of mud, lbs per gallon (L/T)	2.4 (416)	2.4 (416)	1.22 (833)
Base fluid cost (assumed) $/MT ($/L)	1,200 (2.88)	1,200 (2.88)	1,200 (1.45)
Additive weight g/L (kg/ton)	61.3[1] (25)	1 (0.461)	122 (122)
Cost of additive[2] ($/MT)	4,750	40.107[3]	$195[4]
New cost	$5,950	1240[3]	1,395
Percentage Increase in cost	395%	3.3[3]	16.2
Improvement (%)	15	11[3]	77

NOTES:
[1]Dosage was reported at 4 g/L of a 3:1 mix ((LFGO) Large FlakeGO: (PGO) powdered-GO), in terms of carbon content. Carbon content of LFGO was given at 5.2% and that of PGO was given at 30%.
[2]61.3 g/L × 416 L/t.
[3]Data in this publication does not have proper units and limited data is given on the original DF' Fluid Loss volumes. For costing, I calculated the slope of the MWCNT line since it is relatively straight and used that value for 0% addition. I have assumed the mass given as per unit of volume, and using the dosage as in g/L. This evaluation should not be considered as accurate.
[4]Commodity price for rubber.

In Table 7.14, I present an apple to apple comparison for the above discussed works with relevance to their cost to benefit results reported.

Plain elastomer seems to give better results during the fluid loss stage. It is not known how well the material gets displaced in the well completion stage when the formation pressure must force it out.

7.7.1 Environmental motivation

Authorities in most offshore drilling regions are restricting the disposal of oil-based muds, and requiring them to be shipped to shore. The drive for environmentally-friendly DF first resulted in synthetic muds, which had better biodegradability, though these muds are still not allowed in the North Sea. Esters of vegetable oils as DF offer completely biodegradable alternative to oil-based DF. But they have some shortcomings. Chai et al. [520] published one of the more confusing and misleading papers I have read. The title states it is a review of enhancements in ester based DFs, but no discussion of ester based fluids is found in the discussion. They use the term "nanofluids" to describe liquids enhanced by nanomaterials, but dangerously imply the improvements will be the same with ester based DFs. Unfortunately, the enhancement parameters they list citing water and other fluids as surrogates for ester based DFs are far from the primary reasons for failure of low cost ester based fluids. The primary reason for the failure of low carbon alcohol ester-based DF is their chemical makeup. Ester-based DF hydrolyze and decompose due to the double bonds and oxygen content. The breakdown products of this decomposition are carboxylates and low molecular weight alcohols and are toxic to the environment, thus negating (even making worse) the whole point of switching to ester-based natural oils (environmental). Ester hydrolysis immediately starts when the ester comes in contact with water of the invert emulsion systems. Hydrolysis in regular methyl esters of vegetable oils (biodiesel)-based muds is accelerated in temperatures higher than 175°C. 2-Ethyl hexanol is known to be a stable esterification agent, albeit expensive. It would also be logical to use a monocarboxylic ester, i.e., the rest of the carbon chain is saturated. US patent 5,318,956 and [521] outline such a process.

GN may be visualized to stabilize methyl esters. The as-produced GN may be used to form carboxylic groups with the oxygen sites and inhibit hydrolysis. The sites would also help the ester emulsify better with water formulations. Synthesis routes would have to be developed. Palm oil is the cheapest vegetable oil available, so it would be logical to assume palm oil methyl esters would be the target base fluid to modify. A primary issue with palm oil based DFs is the resulting kinematic viscosity. It would be interesting to investigate the resulting kinematic viscosity from impregnation of GN into methyl esters.

7.7.2 Some humble suggestions for applications

Some humble application ideas that may be good topics for research. Except for wastewater treatment, I have not had much experience in the oil & gas industry. As I understand, in older off shore formations, where the (floating) oil level rises beyond the bottom of the production well casing, the seepage of seawater into the well causes large amounts of water to come to the surface with the oil. Being that the oil water interface underground is typically at a high temperature, there is slight emulsification of the oil and separation on surface above becomes a challenge. To avoid this issue, water-soluble polyacrylamides (hydrogel family) are pumped down the well very soon after adding the crosslinking agent. The race to mix and pump is risky, as gelling within the casing could be disastrous. The crosslinked polyacrylamide absorbs water equal to many times its weight, consequently swelling. The hydrophilic swollen polymer then plugs the formation wherever there is water. From the discussions I have had with people over the years, the temperature of the water at that level is sometimes approximately 25−30°C. I have discussed the use of acid washed GN being used as crosslinking agents. If the GN can be functionalized with temperature activated crosslinking agents, such that they change their chemical form at a higher temperature, a premix could be stored at colder temperatures and pumped down without fear of gelling prematurely. This is just a thought, I am not familiar enough to know what other parameters would prevent this application, since drilling muds, completion and production fluids are extremely complicated.

Another problem I am aware of in the production operation is the presence of naphthenic acids. Naphthenic acids are very large molecules of heterocyclic compounds with carboxylic acid groups. The pK_a of these acids will be low, around the pK_a of carboxylic acids, i.e., 4.2. The carboxylic group when protonated has a very high affinity for divalent ions, such as transition and alkaline earth metals. If the formation produces oil with high calcium concentrations, the naphthenic acids will convert to metal naphthalates by ion exchange. The new compound is insoluble and has a specific gravity less than that of water and heavier than that of oil. It therefore forms a layer at the interface that makes separation very difficult, and is then responsible for further scaling due to crystallization. The amount of exchange capacity does not have to be much. There may only be 1 −COOH group for a 14-carbon molecule. But the dissolution product is a very large molecular weight monster. Therefore, even small concentrations of calcium in the produced water can cause large volumes of suspended solids at the oil/water interface. When the concentration of metals is small, we may experiment with GN functionalized with sulfonic groups (SO_3^-). The R-SO_3^- sulfonate group is a strong acid and will tie up the metal ions by ion exchange with the proton. The weak acid (carboxylic) on the naphthenate will get protonated and the precipitation will be avoided. As the GN would be hydrophilic, they could be separated by centrifuge and regenerated by a strong acid such as 5−8% HCl. The concentration of calcium would determine the viability of this application, but being able to reuse the GN would go a long way towards affordability.

SUMMARY AND COMMENTS

I have tried my best in this book to simplify a confusing topic in a holistic manner. It is not a simple topic, so some level of deep dive was necessary. Most people are perplexed when they are in a discussion about nanomaterials. Outside the scientific community, people tend to take the news hype at face value. The high end applications disclosed on the news give the viewer the impression that he/she will soon be driving a car stronger than any material known, or be wearing a shirt as thin as silk that will be bulletproof, monitor his vitals and email any abnormalities directly to his/her doctor, etc. While prototypes of such gadgets may be around, reality is starkly different. With the current state-of-art and the technical and financial hurdles, the hyped products will not be available to anyone not in the 99th percentile. This is precisely why I included many applications that are famously promoted, and detailed the benefits in the section, without prejudice from the commercial end. The cost—benefit analyses that follow such applications should have been unexpected, and hence the resulting reality check is a reminder that all that glitters is not gold.

Graphite has remarkable qualities that can be exploited for a myriad of applications. At the center of all the applications is the reactivity of the material, directly proportional to the quantum of reactive sites. Edge sites of a carbon basal plane (graphene) have dangling bonds that would be bonding them with neighboring carbon atoms with σ bonds. Graphite has plenty of such sites, but due to its uneven surface, the reactive sites seek to lower the energy level and bond with sites on the basal plane, resulting in a low reactivity material.

Nanotubes are like graphene sheets rolled up joining at the edges. The walls are essentially smooth graphene sheets with no defects, and display graphene like characteristics. Some are semiconducting in nature, some are conductors. Unfortunately, the synthesis routes that we currently know of, produce a mix of both types further accompanied by a significant percentage of amorphous carbon. To harness one type over the other from a production batch is a major task and has so far kept nanotubes out of the markets they are uniquely qualified for, such as electronics. Attempts have been made to use nanotubes for applications where the chirality does not matter. Unfortunately, the same quality that makes them candidates for replacement of silicon, ITO, etc. becomes a setback when we try to use them for anything that requires them to react with another material. The walls of the nanotubes, in such instances need to be etched (oxidized) with strong oxidants to create defects on the otherwise smooth surface. The defects create dangling bonds, which make the surface reactive.

GN are made up of uniform size graphene sheets stacked in mainly three configurations with respect to the fiber axis—parallel, at an angle, or perpendicular. The materials are essentially crystalline high purity graphite but with no basal planes exposed to the edge sites of the next graphene sheet. The accessible area is therefore all external, and diffusion limitations are minimal. Since the relatively low energy basal planes are not exposed, and the edge sites are the only sites available, dilute acid treatments can functionalize the edges by forming –COOH and –OH groups. These groups make the fibers hydrophilic and capable of reacting with a multitude of functional groups, which were examined in the functionalization and in-situ polymerization sections. The ease with which GN can be used for reactions makes them a logical choice for such applications. They can be exfoliated to few layer graphene just like graphite, with the huge distinction of producing uniform size graphene platelets instead of uneven size platelets like flake graphite does. They can be attached to polymer matrices, catalysts, crosslinking agents, etc. with predictability because they do not have a complex set of organic oxygenated species on the basal planes.

Flake graphite is currently the material of choice to make graphene oxide (GO), or more appropriately, graphite oxide. The reason, I believe, is the lack of a sustainable supply of GN. Why else would you choose to use harsh chemicals to achieve what GN can deliver with dilute acid treatment? The oxidizing procedure required to make GO from flake graphite has been the same for decades, and produces wastes that are expensive to treat.

Manufacturing GN can be relatively simple. Multiple catalysts and factors such as reaction kinetics, growth rates, catalyst types, and reactor design have been discussed to provide tools for designing a manufacturing process. The rotating type fluidized bed CVD reactors work best for this application, as the catalyst is continually growing fibers, and a turbulent regime to break off the fibers after they reach a certain length cannot be created by gravity type fluidized bed reactors. Catalyst surfaces must be kept clean to avoid fouling as well as expensive purification procedures.

When properly operated, the catalyst particles sit at the end or middle of single fibers and easy to remove with dilute acid wash. When fouled, it is very expensive to purify.

Based on details about manufacturing and purification, a relatively detailed cost analysis for a 1 MT per day capacity plant was performed to put some numbers in perspective. GO was found to be prohibitively expensive for large-scale implementation at $176/kg. The purification process for nanotubes is expensive and results in low yield. A final sale price to industrial users of ~$179/kg was derived. GN production costs were derived from experience. The price to an industrial user would be ~$51/kg.

Finally, a host of applications were examined that look viable with use of GN at first glance. Wherever percentage of nanomaterial dosing was given in literature, the cost addition to the manufacturer was calculated. Some plastics, for example, would have to double in cost to get marginal increases in physical qualities, given the current base price of the material. Some of the applications dropped out based on cost alone. The remaining applications were studied with a holistic approach, always considering practicality (ease) of substitution into current manufacturing practices and realistic expectations for the derived cost-to-benefit ratios. The most viable applications in the near term were determined as follows:

- *Lithium–ion batteries.* The costs for using GN would be about the same as the battery grade graphite being used currently. Physical manufacturing processes do not have to be modified and yet performance enhancement in this application has been proven. The missing link is a sustainable supply of GN.
- *Catalysis.* The costs would be comparable, but higher performance with longer catalyst life can be expected. In exothermic reactions, GN improve temperature control by more efficient heat removal from the catalyst surface and avoid catalyst sintering and subsequent deactivation.
- Additives to improve the physical characteristics of poly lactic acid and other biopolymers. Even a high dose of GN would be affordable by the industries this material is used by. PLA is a high demand polymer but would grow into new, higher value markets if it was stronger and electrically conductive.
- Water treatment:
  - GN are a natural for CDI. They increase the available surface area and react faster to reversing potentials.

- Enhancements of forward osmosis and pervaporation membranes are desperately needed, and can be viable applications right away. Membranes produced from GO, "graphene," etc. are great research topics, but commercially not viable in the near future.
- GN use as catalytic water treatment media, oxidizing H_2S and other dissolved species in situ is a very suitable and real application.

- Gas treatment: GN can be used as a replacement for activated carbon in PSA media for higher performance.
- Lubricants—The growing minimum quantity lubricant market.

Whenever there was an opportunity, I have given some food for thought for students, researchers, and engineers, based on my simple ideas.

FINAL COMMENT: CALL IT GRAPHITE! SELL IT AS GRAPHITE!!

Out of all the applications discussed, the one I feel is the most logical place to start with, and what I tried to explain in vain to my licensee, is simply call GN graphite, sell it against the highest quality graphite. It will certainly perform as well. Consider it a market entry move. As it proves to perform better than competitive products, prices can go up. If we forget about the SEM, TEM and other images at an unimaginable scale, GN is highly crystalline graphite. It can be 100% graphitized at a lower temperature than green graphite. The smaller steel mills, which use an electric arc furnace, consume hundreds of thousands of tons of high quality graphite. They still desire a purity level that is not available in the market. The GN production can be scaled over time, without the customer being afraid of not being able to have a sustainable supply. The target price for this application is well within reach, if realistic business goals are set.

Lastly, I want to say that I have done my best to be accurate in the material I have presented. I am sure there are many errors or signs of ignorance. I welcome comments from anyone to help me be more knowledgeable. The field is vast and material is being published at unprecedented rates. If I were to wait till everything was perfect, the book would never be finished. I hope the material has been helpful.

REFERENCES

[1] D. Kahnemann, Thinking, Fast and Slow, Farrar Straus & Giroux, New York, NY, 2011.
[2] N. Silver, The Signal and the Noise, Penguin Group, UK, 2012.
[3] How the sugar industry shifted blame to fat. NY Times, September 12, 2016.
[4] A. Chambers, C. Park, R.T.K. Baker, N. Rodriguez, Hydrogen storage in graphite nanofibers, J. Phys. Chem. B. 102 (22) (1998) 4253–4256.
[5] M. Rzepka, E. Bauer, G. Reichenauer, T. Schlierman, B. Bernhardt, K. Bohmhammel, et al., Hydrogen storage capacity of catalytically grown carbon nanofibers, J. Phys. Chem. B. 109 (31) (2005) 14979–14989.
[6] C. Shan, H. Yang, D. Han, Q. Zhang, A. Ivaska, L. Niu, Water-soluble graphene covalently functionalized by biocompatible poly-L-lysine, Langmuir. 25 (20) (2009) 12030–12033.
[7] M. Kole, T.K. Dey, Investigation of thermal conductivity, viscosity, and electrical conductivity of graphene based nanofluids, J. Appl. Phys. 113 (2013) 084307.
[8] C. Yan, Y.W. Kanaththage, R. Short, C.T. Gibson, L. Zou, Graphene/polyaniline nanocomposite as electrode material for membrane capacitive deionization, Desalination 344 (2014) 274–279.
[9] S. Stankovich, D.A. Dikin, G.H.B. Dommett, M.K. Kevin, E.J. Zimney, E.A. Stach, et al., Graphene based composite materials, Nature 442 (2006) 282–286.
[10] J. Samuel, J. Rafiee, P. Dhiman, Z. Yu, N. Koratkar, Graphene colloidal suspensions as high performance semi-synthetic metal-working fluids, J. Phys. Chem. 115 (2011) 3410–3415.
[11] K. Nakada, M. Fujita, G. Dresselhaus, M.S. Dresselhaus, Edge state in grapheme ribbons: nanometer size effect and edge shape dependence, Phys. Rev. B 54 (1996) 17954–17961.
[12] Y. Liu, A. Dobrinsky, B.I. Yakobson, Graphene edge from A to Z—and the origins of nanotube chirality, Phys. Rev. Lett. 105 (2010) 235502.
[13] E. Toshiak, Graphene edges and nanographene—electronic structure and nanofabrications, in: Tokyo Inst of Tech. Joint Symp, Turkey, March 18–19, 2010.
[14] K. Katsuyoshi, Electronic structure of a stepped graphite surface, Phys. Rev. B 48 (1993) 1757.
[15] L.A. Girifalco, M. Hodak, S.L. Lee, Carbon nanotubes, buckyballs, ropes, and a universal graphitic potential, Phys. Rev. B 62 (2000) 13104.
[16] M. Endo, R. Saito, From Carbon Fibers to Nanotubes, CRC Press, New York, NY, 1997.
[17] D.H. Robertson, D.W. Brenner, C.T. White, On the way to fullerenes: molecular dynamics study of the curling and closure of graphitic ribbons, J. Phys. Chem. 96 (15) (1992) 6133–6135.
[18] A. Peigney, P. Coquay, E. Flahaut, R.E. Vandenbergghe, E.D. Grave, C. Laurent, A study of the formation of single- and double-walled carbon nanotubes by a CVD method, J. Phys. Chem. B 105 (40) (2001) 9699–9710.
[19] D. Jiang, B.G. Sumpter, S. Dai, The unique chemical reactivity of a graphene nanoribbon's zigzag edge, J. Chem. Phys. 126 (2007) 134701.
[20] P.E. Anderson, A method for characterization and quantification of platelet graphite nanofiber edge crystal structure, Carbon 44 (2006) 2184–2190.

[21] S.E. Stein, R.L. Brown, Prediction of carbon-hydrogen bond dissociation energies for polycyclic aromatic hydrocarbons of arbitrary size, J. Am. Chem. Soc. 113 (3) (1991) 787–793.
[22] S.V. Rotkin, Y. Gogotsi, Analysis of non-planar graphitic structures: from arched edge planes of graphite crystals to nanotubes., Mater. Res. Innov. 5 (2000) 191–200.
[23] S.V. Rotkin, R.A. Suris, Bond passivation model: diagram of carbon nanoparticles, Phys. Lett. A 261 (1999) 98–101.
[24] S.V. Rotkin, On energetics of NT nucleation through zipping of carbon layer edge. PV 2000-12 in: P.V. Kamat, D.M. Guldi, K.M. Kadish (Eds.), Fullerenes 2000, vol. 10: Chemistry and Physics of Fullerenes and Carbon Nanomaterials, ECS Inc., Pennington, NJ, 2000, pp. 66–71.
[25] S.V. Rotkin, R.A. Suris, Carbon cluster formation energy, in: R.S. Ruoff, M. Kadish (Eds.), Fullerenes. Recent Advances in the Chemistry and Physics of Fullerenes and Related Materials, vol. II, Electrochemical Society, Pennington, NJ, 1995, pp. 1263–1270. PV 10-95.
[26] D.H. Robertson, D.W. Brenner, J.W. Mintmire, Energetics of nanoscale graphitic tubules, Phys. Rev. B 45 (1992) 12592.
[27] M.B. Nardelli, B.I. Yakobson, J. Bernholc, Mechanism of strain release in carbon nanotubes, Phys. Rev. B 57 (1998) R4277–R4280.
[28] G.A. Korn, T.M. Korn, Mathematical Handbook for Scientists and Engineers, Mcgraw-Hill, New York, NY, 1968.
[29] L.X. Benedict, N.G. Chopra, M.L. Cohen, A. Zetel, S.G. Louie, V.H. Crespi, Microscopic determination of the interlayer binding energy in graphite, Chem. Phys. Lett. 286 (1998) 490–496.
[30] E.S. Oh, Van der Waals interaction energies between non-planar bodies Korean, J. Chem. Eng. 21 (2) (2004) 494–503.
[31] A.I. Zhbanov, E.G. Pogorelov, Y. Chang, Van der Waals interaction between two crossed carbon nanotubes, Condens. Matter Mater. Sci (2008).
[32] S.N. Thennadil, L.H. Garcia-Rubio, Approximations for calculating van der Waals interaction energy between spherical particles—a comparison, J. Colloid Interface Sci. 243 (2001) 136–142.
[33] J.W. Ager III, Residual stress in diamond and amorphous carbon films, Mater. Res. Soc. Symp. Proc. 383 (1995) 143–151.
[34] G.S. Wulff, Wulff construction and first principles, Z. Kristallogr. Min. 34 (1901) 449–530.
[35] S.V. Rotkin, I. Zharov, K. Hess, Zipping of graphene edge results in [10, 10] (TIFF), in: H. Kuzmany, J. Fink, M. Mehring, S. Roth (Eds.), Electronic Properties of Molecular Nanostructures; (XVth International Winterschool/Euroconference, Kirchberg, Tirol, Austria, 3–10 March 2001), AIP Conference Proceedings, vol. 591, 2001, pp. 454–457.
[36] L. Li, S. Reich, J. Robertson, Defect energies of graphite: density-functional calculations, Phys. Rev. B 72 (2005) 184109.
[37] R. Zacharia, Oxidation of graphitic surfaces (Dissertation online), Freie Universität Berlin, May 2004 (Chapter 5).
[38] R.T. Yang, C. Wong, Mechanism of single-layer graphite oxidation: evaluation by electron microscopy, Science 214 (4519) (1981) 437–438.
[39] Z. Klusek, P.K. Datta, W. Kozlowski, Nanoscale studies of the oxidation and hydrogenation of graphite surface, Corros. Sci. 45 (2003) 1383–1393.
[40] R.C. Bansal, F.J. Vastola, P.L. Walker Jr, Studies on ultraclean carbon surfaces-III. Kinetics of chemisorption of hydrogen on graphon, Carbon 9 (2) (1971) 185–192.
[41] P. Ehrburger, C. Vix-Guterl, Proc. Eurocarbon, France, 1998.

[42] M.R. Lopez, T.J. Lee, M. Bella, C. Salzman, Formation and Chemistry of Carboxylic Anhydrides at the Graphene Edge, The Royal Society of Chemistry, London, 2013, pp. 1–3.
[43] D. Rosenthal, M. Ruta, R. Schlo, L.K. Minsker, Combined XPS and TPD study of oxygen-functionalized carbon nanofibers grown on sintered metal fibers, Carbon 48 (2010) 1835–1843.
[44] G.J. Gleicher, Hydrogen abstraction from substituted benzyl chlorides by the trichloromethyl radical, Tetrahedron (8) (1974) 935–938.
[45] J.S. Bunch, S.S. Verbridge, J.S. Alden, A.M. Van der Zande, J.M. Parpia, H.G. Craighead, et al., Impermeable atomic membranes from graphene sheets, Nano Lett. 8 (8) (2008) 2458–2462.
[46] K.S. Novoselov, A.K. Geim, S.V. Morozov, D. Jiang, Y. Zhang, D.S.V. Grigorieva IV, et al., Electric field effect in atomically thin carbon films, Science 306 (5696) (2004) 666–669.
[47] A.C.G. Arramel, B. Jan van Wees, Band gap opening of graphene by noncovalent π–π interaction with porphyrins, Graphene 2 (2013) 102–108.
[48] D. Jahani, Electronic tunneling in graphene, in: J.R. Gong (Ed.), New Progress on Graphene Research, InTech, (2013) http://dx.doi.org/10.5772/51980. Available from: http://www.intechopen.com/books/new-progress-on-graphene-research/electronic-tunneling-in-graphene.
[49] R. Britnell, V. Gorbachev, R. Jalil, B.D. Belle, F. Schedin, A. Mishchenko, et al., Field-effect tunneling transistor based on vertical graphene heterostructures, Science 335 (6071) (2012) 947–950.
[50] A. Kargar, D. Wang, Analytical modeling of graphene nanoribbon Schottky diodes, in: Proc. of SPIE, vol. 7761.
[51] A.K. Geim, K.S. Novoselov, Rise of graphene, Nat. Mater. 6 (2007) 183–191.
[52] P. Avouris, Graphene: electronic and photonic properties and devices, Nano Lett. 10 (11) (2010) 4285–4294.
[53] Z. Wei, D. Wang, S. Kim, S.Y. Kim, Y. Hu, M.K. Yakes, et al., Nanoscale tunable reduction of graphene oxide for graphene electronics, Science 328 (5984) (2010) 1373–1376.
[54] X. Li, G. Zhang, X. Bai, X. Sun, X. Wang, E. Wang, et al., Highly conducting graphene sheets and Langmuir–Blodgett films, Nat. Nanotechnol. 3 (2008) 538–542.
[55] A.B. Silva-Tapia, F.V. Burgos, L.R. Radovic, Oxygen migration on the graphene surface. 1. Origin of epoxide groups, Carbon 49 (2011) 4218–4225.
[56] J.M. Baranowski, D.P. Mozdonek, K. Grodecki, P. Osewski, W. Kozlowski, M. Kopciuszynski, et al., Observation of electron-phonon couplings and fano resonances in epitaxial bilayer graphene, Graphene 2 (4) (2013) 115–120.
[57] A.S. David, Electronic structure of single layer graphene (Ph.D. Thesis), UC Berkeley, 2002.
[58] S.D. Sarma, S. Adam, E.H. Hwang, E. Rossi, Electronic transport in two-dimensional graphene, Rev. Mod. Phys 83 (2011) 407.
[59] D.R. Cooper, B. D'Anjou, N. Ghattamaneni, B. Harack, M. Hilke, A. Horth, et al., Experimental review of graphene, Condens. Matter Phys. 2012 (2012). Article ID 501686.
[60] A.F. Young, Y. Zhang, P. Kim, Experimental manifestation of berry phase in graphene, Physics of Graphene, NanoScience and Technology, Springer, Switzerland, 2014. Available from: http://dx.doi.org/10.1007/978-3-319-02633-6-1.
[61] Z. Zhang, T. Li, Graphene morphology regulated by nanowire patterned in parallel on a substrate surface, J. Appl. Phys. 107 (2010) 103519. Also selected to be included in Virtual J. Nano Sci. Tech., Vol. 21, Issue 23 (2010).

[62] F. Banhart, J. Kotakoski, A.V. Krasheninnlkov, Structural defects in graphene, ACS Nano 5 (1) (2011) 26–41.
[63] G.C. Rodriguez, P. Zelenovskiy, K. Romanyuk, S. Luchkin, Y. Kopelevich, A. Kholkin, Strong piezo electric in single layer graphene deposited on SiO_2 grating, Nat. Commun. 6 (2015). Article No. 7572.
[64] A. Balandin, S. Ghosh, W. Bao, I. Calizo, D. Teweldebrhan, F. Miao, et al., Superior thermal conductivity of single layer graphene, Nano Lett. 8 (3) (2008) 902–907.
[65] N.L. Ritzert, W. Li, C. Tan, G.G. Rodriguez-Calero, J. Rodrigues-Lopez, K. Burgos, et al., Single layer graphene as an electrochemical platform, Faraday Discuss. 172 (2014) 27–45.
[66] Lai, K.W.C., Xi, N., Chen, H., Fung, C.K.M., & Chen, L., Zero-bandgap graphene for infrared sensing applications, in: SPIE Defense, Security, and Sensing, International Society for Optics and Photonics, (May 2011), pp. 80312N–80312N.
[67] P.Y. Huang, C. Ruiz-Vargas, A.M. Van der Zande, W.S. Whitney, M.P. Levendorf, J.W. Kevek, et al., Grains and grain boundaries in single-layer graphene atomic patchwork quilts, Nature 469 (2011) 389–392.
[68] Z.S. Wu, S. Pei, W. Ren, D. Tang, L. Gao, B. Liu, et al., Field emission of single-layer graphene films prepared by electrophoretic deposition, Adv. Mater. 21 (17) (2009) 1756–1760.
[69] L. Campos, V.R. Manfrinato, J.D.S. Yamagishi, J. Kong, P.J. Herrero, Anistropic etching and nanoribbon formation in single layer graphene, Nano Lett. 9 (7) (2009) 2600–2604.
[70] F. Scarpa, S. Adhikari, A.S. Phani, Effective elastic mechanical properties of graphene, Nanotechnology. 20 (6) (2009) 065709.
[71] A.V. Shytov, M.I. Katsnelson, L.S. Levitov, Atomic collapse and quasi–Rydberg states in graphene, Phys. Rev. Lett 99 (2007) 246802.
[72] J.W. Colson, A.R. Woll, A. Mukherjee, M.P. Levendorf, E.L. Spitler, V.B. Shields, et al., Oriented 2D covalent organic framework thin films on single-layer graphene, Science 332 (6026) (2011) 228–231.
[73] H. Wei, K. Pi, K.M. McCreary, Y. Li, J.J.I. Wong, A.G. Swartz, et al., Tunneling spin injection into single layer graphene, Phys. Rev. Lett. 105 (2010) 167202.
[74] A. Kargar, Analytical modeling of graphene nanoribbon Schottky diodes, Proc. of SPIE, vol. 7761, 2010.
[75] X. Zang, Q. Zhou, J. Chang, Y. Liu, L. Liwei, Graphene and carbon nanotube (CNT) in MEMS/NEMS applications, Microelectron. Eng. 132 (2015) 192–206.
[76] S. Rumyantsev, G. Liu, W. Stillman, M. Shur, A.A. Balandin, Electrical and noise characteristics of graphene field-effect transistors: ambient effects, noise sources and physical mechanisms, J. Phys. Condens. Matter 22 (39) (2010).
[77] J.T. Robinson, M. Zalalutdinov, J.W. Baldwin, E.S. Snow, Z. Wei, P. Sheehan, et al., Wafer-scale reduced graphene oxide films for nanomechanical devices, Nano Lett. 8 (10) (2008) 3441.
[78] J.S. Bunch, A.M. Van der Zande, S.S. Verbridge, I.W. Frank, D.M. Tanenbaum, J.M. Parpia, et al., Electromechanical resonators from graphene sheets, Science 315 (5811) (2007) 490–493.
[79] R.R. Nair, A.N. Grigorenko, K.S. Novoselov, T.J. Booth, R. Peres, A.K. Geim, Fine structure constant defines visual transparency of graphene, Science 320 (5881) (06 Jun 2008) 1308.
[80] J. Wu, H.A. Becerril, Z. Bao, Z. Liu, Y. Chen, P. Peumans, Organic solar cells with solution processed graphene transparent electrodes, Appl. Phys. Let. 92 (2008) 26.
[81] K.S. Kim, Y. Zhao, H. Jang, S.Y. Lee, J.M. Kim, K.S. Kim, et al., Large-scale pattern growth of graphene films for stretchable transparent electrodes, Nature. 457 (2009) 706–710.

[82] X. Wang, L. Zhi, K. Müllen, Transparent, conductive graphene electrodes for dye-sensitized solar cells, Nano Lett. 8 (2008) 323.
[83] H. Valera-Rizo, I. Martin-Gullon, M. Terrones, Hybrid films with graphene oxide and metal nanoparticles could now replace indium tin oxide, ACS Nano 6 (6) (2012) 4565–4572.
[84] I. Vlassiouk, G. Polizos, R. Cooper, I. Ivanov, J.K. Keum, P. Datskos, et al., Strong and electrically conductive graphene-based composite fibers and laminates, ACS Appl. Mater. Interfaces 7 (20) (2015) 10702–10709.
[85] N.W. Pu, C.A. Wang, Y. Sung, Y.M. Liu, M.D. Ger, Production of few-layer graphene by supercritical CO_2 exfoliation of graphite, Mater. Lett. 63 (2009) 1987–1989.
[86] E. Ewidenkvist, D.W. Boukhvalov, S. Rubino, S. Akhtar, J. Lu, R.A. Quinlan, et al., Mild sonochemical exfoliation of bromine-intercalated graphite: a new route towards graphene, J. Phys. D: Appl. Phys 42 (2009) 112003.
[87] Y. Sim, P. Park, Y.J. Kim, M.J. Seong, Synthesis of graphene layers using graphite dispersion in aqueous surfactant solutions, J. Korean Phys. Soc. 58 (4) (2011) 938–942.
[88] X. Chen, L. Zhang, S. Chen, Large area CVD growth of graphene, Synth. Met. (2015).
[89] Y. Lee, S. Bae, H. Jang, S. Jang, S.E. Zhu, S.H. Sim, et al., Wafer-scale synthesis and transfer of graphene films, Nano Lett. 10 (2) (2010) 490–493.
[90] T.D. Shen, W.Q. Ge, K.Y. Wang, M.X. Quan, J.T. Wang, W.D. Wei, et al., Structural disorder and phase transformation in graphite produced by ball milling, Nanostruct. Mater. 7 (4) (1996) 393–399.
[91] Y. Chen, J. Fitzgerald, L.T. Chadderton, L. Chaffron, Nanoporous carbon produced by ball milling, Appl. Phys. Lett. 74 (1999) 19.
[92] Y. Hernandez, V. Nicolos, M. Lotya, F.M. Blighe, Z. Sun, S. Sukanta De, High-yield production of graphene by liquid-phase exfoliation of graphite, Nat. Nanotechnol. 3 (2008) 563–568.
[93] C. Fu, X. Yang, Molecular simulation of interfacial mechanics for solvent exfoliation of graphene from graphite, Carbon 55 (2013) 350–360.
[94] S.H. Anderson, D.D.L. Chung, Exfoliation of single crystal graphite and graphite fibers intercalated with halogens, Synthetic Mater. 8 (1983) 343–349.
[95] C.A. Martin, J.K.W. Sandler, A.H. Windle, M.K. Schwarz, W. Bauhofer, K. Schulte, et al., Electric field-induced aligned multi-wall carbon nanotube networks in epoxy composites, Polymer 46 (2005) 877–886.
[96] P. Glatkowski, P. Mack, J.L. Conroy, J.W. Piche, P. Winsor, Electromagnetic shielding composite comprising nanotubes, US Patent 6,265,466, 2001.
[97] J.H. Li, G.S. Kim, Y.M. Choi, W. Park, J.A. Rogers, U. Paik, Comparison of multiwalled carbon nanotubes and carbon black as percolative paths in aqueous-based natural graphite negative electrodes with high-rate capability for lithium-ion batteries, J. Power Sources 184 (2008) 308–311.
[98] J.P. Hill, W. Jin, A. Kosaka, T. Fukushima, H. Ichihara, T. Shimomura, et al., Self-assembled hexa-peri-hexabenzocoronene graphitic nanotube, Science 304 (2004).
[99] R.E. Camacho, A.R. Morgan, M.C. Flores, T.A. McLeod, V.S. Kumsomboone, B.J. Mordecai, et al., Carbon nanotube arrays for photovoltaic applications, J. Miner. Metals Mater. Soc. 59 (3) (March 2007) 39–42.
[100] T.K. Manna, S.M. Mahajan, Nanotechnology in the development of photovoltaic cells, in: Int'l. Conf. on Clean Electrical Power, 2007, pp. 379–386.
[101] Du.C. Chunsheng, J. Yeh, N. Pan, High power density supercapacitors using locally aligned carbon nanotube electrodes, Nanotechnology. 16 (4) (2005).
[102] J.M. Bonard, M. Croci, C. Klinke, R. Kurt, O. Noury, N. Weiss, Carbon nanotube films as electron field emitters, Carbon 40 (10) (2002) 1715–1728.

[103] X.P. Gao, Y. Zhang, G.L. Pan, J. Yan, F. Wu, H.T. Yuan, et al., Carbon nanotubes filled with metallic nanowires, Carbon 42 (1) (2004) 47–52.
[104] J.U. Li, Photovoltaic effect in ideal carbon nanotube diodes, Appl. Phys. Lett. 87 (2005) 073101.
[105] D.T. Welna, L. Qu, B.E. Taylor, L. Dai, M.F. Durstock, Vertically aligned carbon nanotube electrodes for lithium-ion batteries, J. Power Sources 196 (2011) 1455–1460.
[106] U.J. Kim, X.M. Liu, C.A. Furtado, G. Chen, R. Saito, J. Jiang, et al., Infrared-active vibrational modes of single-walled carbon nanotubes, Phys. Rev. Lett. 95 (2005) 157402.
[107] S. Trasobares, C.P. Ewels, J. Birrell, O. Stephan, B.Q. Wei, J.A. Carlisle, et al., Carbon nanotubes with graphitic wings, Adv. Mater. 16 (7) (2004).
[108] Q. Li, Z. Ni, J. Gong, D. Zhu, Z. Zhu, Carbon nanotubes coated by carbon nanoparticles of turbostratic stacked graphenes, Carbon 46 (3) (2008) 434–439.
[109] J.M.C. Moreno, M. Yoshimura, Hydrothermal processing of high-quality multiwall nanotubes from amorphous carbon, J. Am. Chem. Soc. 123 (2001) 741–742.
[110] D.R. Dreyer, S. Park, C.W. Bielawski, R.S. Ruoff, The chemistry of graphene oxide, Chem. Soc. Rev. 39 (2010) 228–240.
[111] B.C. Brodie, On the atomic weight of graphite, Philos. Trans. R. Soc. Lond. 149 (1859) 249–259.
[112] L. Staudenmaier, Verfahren zur darstellung der Graphitsäure (Procedure for the presentation of graphite oxide), Ber. Dtsch. Chem. Ges. 31 (1898) 1481–1487.
[113] W.S. Hummers, R.E. Offeman, Preparation of graphitic oxide, J. Am. Chem. Soc. 80 (1958) 1339.
[114] M.R. Lopez, C.G. Salzmann, A simple and mild chemical oxidation route to high-purity nano-graphene oxide, Carbon 106 (2016) 56–63.
[115] T. Rattana, S. Chaiakun, N. Witit-anun, N. Nuntawong, P. Chindaudom, S. Oaew, et al., Preparation and characterization of graphene oxide nanosheets, Proc. Eng. 32 (2012) 759–764.
[116] K. Zhang, H. Li, X. Xu, H. Yu, Facile and efficient synthesis of nitrogen functionalized graphene oxide as a copper adsorbent and its application, Ind. Eng. Chem. Res. 55 (8) (2016) 2328–2335.
[117] S. Abdolhosseinzadeh, H. Asgharzadeh, S.H. Kim, Fast and fully-scalable synthesis of reduced graphene oxide, Sci. Rep. 5 (2015) 10160.
[118] F.Y. Ban, S.R. Majid, N.M. Huang, H.N. Lim, Graphene oxide and its electrochemical performance, Int. J. Electrochem. Sci. 7 (2012) 4345–4351.
[119] S. Stankovich, R.D. Piner, S.T. Nguyen, R.S. Ruoff, Synthesis and exfoliation of isocyanate treated graphene oxide nanoplatelets, Carbon 44 (15) (2006) 3342–3347.
[120] D.C. Marcan, D.V. Kosynkin, J.M. Berlin, A. Sinitskli, Z. Sun, A. Slesarev, et al., Improved synthesis of graphene oxide, ACS Nano 4 (8) (2010) 4806–4814.
[121] M. Fathy, A. Gomaa, F.A. Taher, M.M. El-Fass, A.E.H.B. Kashyout, Optimizing the preparation parameters of GO and rGO for large-scale production, J. Mater. Sci. 51 (12) (2016) 5664–5675.
[122] H. Xiang, K. Zhang, G. Ji, J.Y. Lee, C. Zou, X. Chen, et al., Graphene/nanosized silicon composites for lithium battery anodes with improved cycling stability, Carbon 49 (5) (2011) 1787–1796.
[123] C. Park, R.T. Baker, Catalytic behavior of graphite nanofiber supported nickel particles. 3. The effect of chemical blocking on the performance of the system, J. Phys. Chem. B. 103 (1999) 2453.
[124] T.G. Ros, D.E. Keller, A.J. van Dillen, J.W. Geus, Koningsberger, Preparation and activity of small rhodium metal particles on fishbone carbon nanofibres, J. Catal. 211 (2002) 85.

[125] A. Lucas, Ph Lambin, R.E. Smalley, On the energetics of tubular fullerenes, J. Phys. Chem. Solids. 54 (1993) 587–593.
[126] J.C. Charlier, A. De Vita, X. Blasé, R. Car, Microscopic growth mechanisms for carbon nanotubes, Science 275 (1997) 646–649.
[127] Y.K. Kwon, Y.H. Lee, S.G. Kim, P. Jund, D. Tománek, R.E. Smalley, Morphology and stability of growing nanotubes, Phys. Rev. Lett. 79 (1997) 2065.
[128] V.R. Choudhary, A.S. Mamman, Energy efficient conversion of methane to syngas over NiO-MgO solid solution, Appl. Energy 66 (2000) 161–175.
[129] Y. Echegoyen, I. Suelves, M.J. Lazaro, M.L. Sanjuan, R. Moliner, Thermo catalytic decomposition of methane over Ni–Mg and Ni–Cu–Mg catalysts: effect of catalyst preparation method, Appl. Catal. A 333 (2007) 229–237.
[130] D. Hulicova-Jurcakova, X. Li, Z. Zhu, Z. de Marco, G.Q. Lu, Graphitic carbon nanofibers synthesized by the chemical vapor deposition (CVD) method and their electrochemical performances in supercapacitors, Energy Fuels. 22 (6) (2008) 4139–4145.
[131] M.A. Ermakova, D. Ermakov, G.G. Kuvshinov, L.M. Plyasova, New Ni catalysts for the formation of filamentous carbon in the reaction of methane decomposition, J. Catal. 187 (1999) 77–84.
[132] T.V. Reshetenko, L.B. Avdeeva, Z.R. Ismagilov, A.L. Chuvilin, V.A. Ushakov, Carbon capacious Ni-Cu-Al_2O_3 catalysts for high temperature methane decomposition, Appl. Catal. A 247 (2003) 51–63.
[133] S. Takenaka, S. Kobayashi, H. Ogihara, K. Otsuka, Ni/SiO_2 catalyst effective for methane decomposition into hydrogen and carbon nanofibers, J. Catal. 217 (2003) 79–87.
[134] X. Li, Y. Zhang, K. Smith, Metal–support interaction effects on the growth of filamentous carbon over Co/SiO_2 catalysts, Appl. Catal. A 264 (2004) 81–91.
[135] J. Ashok, P.S. Reddy, G. Raju, M. Subrahmanyam, A. Venugopal, Catalytic decomposition of methane to hydrogen and carbon nanofibers over Ni-Cu-SiO_2 catalysts, Energy Fuels 23 (2009) 5–13.
[136] A.A. Puretzky, G.S. Jesse, I.N. Ivanov, G. Eres, In situ measurements and modeling of carbon nanotube array growth kinetics during chemical vapor deposition, Appl. Phys. A 81 (2005) 223–240.
[137] W.R. Davis, R.J. Slawson, G.R. Rigby, An unusual form of carbon, Nature. 171 (1953) 756.
[138] H. Terrones, T. Hayashi, M.M. Navia, M. Terrones, Y. Kim, M.E. Grobert, et al., Graphitic cones in palladium catalyzed carbon nanofibers, Chem. Phys. Lett. 343 (2001) 241–250.
[139] H. Murayama, T. Maeda, A novel form of filamentous carbon, Nature. 345 (1990) 791–793.
[140] A. Peigney, P. Coquay, E. Flahaut, R.E. Vandenberghe, E. De Grave, A. Laurent, Study of the formation of single- and double-walled carbon nanotubes by a CVD method, J. Phys. Chem. B. 105 (2001) 9699–9710.
[141] A.V. Krestinin, Formation of soot particles as a process involving chemical condensation of polyines, Chem. Phys. Rep. c/c of Khimicheskaia Fizika 17 (8) (1998) 1441–1462.
[142] K. Zhenhui, E. Wang, L. Gao, S. Lian, M. Jiang, C. Hu, et al., One-step water-assisted synthesis of high-quality carbon nanotubes directly from graphite, J. Am. Chem. Soc. 125 (45) (2003) 13652–13653.
[143] L.M. Viculis, J.J. Mack, R.B. Kaner, A chemical route to carbon nanoscrolls, Science 299 (2003) 1361.
[144] K. Fujisawa, T. Hasegawa, D. Shimamoto, H. Muramatsu, Y.C. Jung, T. Hayashi, et al., Boron atoms as loop accelerator and surface stabilizer in platelet type carbon nanofibers, Chem. Phys. Chem 11 (2010) 2345–2348.

[145] J. Campos-Delgado, J.M. Romo-Herrera, X. Jia, D.A. Cullen, H. Muramatsu, Y.A. Kim, et al., Bulk production of a new form of sp2 carbon: crystalline graphene nanoribbons, J. Nano Lett. 8 (2008) 2773–2778.
[146] S.H. Yoon, S. Lim, S. Hong, W. Qiao, D.D. Whitehurst, I. Mochida, et al., Conceptual model for the structure of catalytically grown carbon nano-fibers, Carbon 43 (2005) 1828–1838.
[147] C.A. Bernardo, J.R. Rostrup Nielsen, Carbon deposition and methane steam reforming on silica supported Ni-Cu catalysts, J. Catal. 96 (1985) 517–534.
[148] R.T.K. Baker, M.A. Barber, P.A. Harris, F.S. Feates, R.J. Waite, Nucleation and growth of carbon deposits from the nickel catalyzed decomposition of acetylene, J. Catal. 261 (1972) 51–62.
[149] J. Rostrup-Nielson, D.L. Trimm, Mechanism of carbon formation on nickel containing catalysts, J. Catal. 48 (1977) 155.
[150] M. Boudart, W.L. Holstein, The temperature difference between a supported catalyst particle and its support during exothermic and endothermic catalytic reaction, Rev. Latino-am. Ing. Quim. Apl 13 (1983) 107.
[151] P.K. De Bokx, A.J.H.M. Kock, E. Boellaard, W. Klop, J.W. Geus, The formation of filementaous carbon on iron and nickel catalysts. Part 1, J. Catal. 96 (2) (1985) 454.
[152] P.K. De Bokx, A.J.H.M. Kock, E. Boellaard, W. Klop, J.W. Geus, The formation of filementaous carbon on iron and nickel catalysts. Part 2. Mechanism, J. Catal. 96 (2) (1985) 468.
[153] P.K. De Bokx, A.J.H.M. Kock, E. Boellaard, W. Klop, J.W. Geus, The formation of filementaous carbon on iron and nickel catalysts. Part 3. Morphology, J. Catal. 96 (2) (1985) 481–490.
[154] J.W. Snoeck, G.F. Froment, M. Fowles, Filamentous carbon formation and gasification: thermodynamics, driving force, nucleation and steady state growth, J. Catal 169 (1) (1997) 240–249.
[155] S. Helveg, C. López-Cartes, J. Sehested, P.L. Hansen, B.S. Clausen, J.R. Rostrup-Nielsen, et al., Atomic-scale imaging of carbon nanofibre growth, Nature. 427 (2004) 426–429.
[156] P. Ammendola, R. Chirone, G. Ruoppolo, G. Russo, Production of hydrogen from thermo-catalytic decomposition of methane in a fluidized bed reactor, Chem. Eng. J. 154 (2009) 287–294.
[157] L.M. Das, The potential for hydrogen as an alternative fuel for transportation and power sector in India, Int. J. Environ. Stud. 64 (6) (2007) 749–759.
[158] J. Jangbarwala, Final report on LFG to GNF at Bonsall landfill, SD County, CA, (2009), Catalyx, Inc., info@catalyxinc.com.
[159] M.A. Van der Hoef, M. van Sint Annaland, N.G. Deen, J.A.M. Kuipers, Numerical simulation of dense gas-solid fluidized beds: a multiscale modeling strategy, Annu. Rev. Fluid Mech. 40 (2008) 47–70.
[160] R. Fan, R.O. Fox, M.E. Muhle, Role of intrinsic kinetics and catalyst particle size distribution in CFD simulations of polymerization reactors, in: The 12th International Conference on Fluidization, May 13–17, Vancouver, Canada, (2007), 993–1000.
[161] T. Mikami, H. Kamiya, M. Horio, Numerical simulation of cohesive powder behavior in a fluidized bed, Chem. Eng. Sci. 53 (10) (1998) 1927–1940.
[162] J. Wang, Y. Liu, EMMS-based Eulerian simulation on the hydrodynamics of a bubbling fluidized bed with FCC particles, Powder Tech. 197 (3) (2010) 241–246.
[163] W. Wu, S.A. Jayarathna, B.M. Halvorsen, Experimental study of effects of particle size distribution on bubble behavior for validation of CFD modeling of bubbling fluidized bed, Small 400 (2008) 600.

[164] S. Das Sharma, T.S. Pugsley, Effect of particle size distribution on the performance of a fluidized bed reactor, in: The 12th International Conference on Fluidization, 2007.
[165] M.A. Van der Hoef, R. Beetstra, J.A.M. Kuipers, Lattice-Boltzmann simulations of low-Reynolds-number flow past mono- and bidisperse arrays of spheres: results for the permeability and drag force, J. Fluid. Mech. 528 (2005) 233–254.
[166] R. Fan, D.L. Marchiso, R.O. Fox, Application of the direct quadrature method of moments to polydisperse gas–solid fluidized beds, Powder Technol. 139 (1) (2004) 7–20.
[167] M. Andrews, P. O'Rourke, The multiple particle-in-cel (MP-PIC) method for dense particle flow, Int. J. Multiph. Flow 22 (1996) 379–402.
[168] N.A. Patankar, D.D. Joseph, Modeling and numerical simulation of particulate flows by the Eulerian–Lagrangian approach, Int. J. Multiph. Flow 27 (10) (1996) 379–402.
[169] W. Wang, S. Dai, X. Lim, J. Yang, D.J. Srolovitz, Q. Zheng, Measurement of the cleavage energy of graphite, Nat. Commun. 6 (2015). Article No. 7853.
[170] L.A. Girifalco, Molecular properties of fullerene in the gas and solid phases, J. Phys. Chem. 96 (1992) 858.
[171] A.I. Zhbanov, G. Pogorelov, Y.C. Chang, Van der Waals interaction between two crossed carbon nanotubes, ACS Nano 4 (2010) 5937–5945.
[172] M. Sidlecki, W. Jong, Verkooijen, Fluidized bed gasification as a mature and reliable technology for the production of bio-syngas and applied in the production of liquid transportation fuels—a review, Energies 4 (2011) 389–434.
[173] Y.M. Chen, Fundamentals of a centrifugal fluidized bed, AIChE J. 33 (5) (1987) 722–728.
[174] L.A. Anderson, S.H. Hasinger, B.N. Turman, Two-component vortex flow studies of the colloid core nuclear rocket, J. Spacecr. Rockets. 9 (5) (1972).
[175] E.P. Volchkov, V.I. Terekhov, A.N. Kaidanik, A.N. Yadkin, Aerodynamics and heat and mass transfer of fluidized particles beds in vortex chambers, Heat Transfer Eng. 14 (3) (1993) 36–47.
[176] M. Goldshtik, F. Hussain, R.J. Yao, The vortex liquid piston engine and some other vortex technologies, Sadhana 22 (3) (1997) 323–367.
[177] A. De Broqueville, J. De Wilde, Numerical investigation of gas–solid heat transfer in rotating fluidized beds in a static geometry, Chem. Eng. Sci. 64 (6) (2009) 1232–1248.
[178] J. De Wilde, A. de Broqueville, Rotating fluidized beds in a static geometry:experimental proof of concept, AIChE J. 53 (4) (2007) 793–810.
[179] J. De Wilde, A. de Broqueville, Experimental investigation of a rotating fluidized bed in a static geometry, Powder Technol. 183 (2008) 426–435.
[180] A.R. Murray, E.R. Kisin, A.V. Tkach, N. Yanamala, R. Mercer, S.H. Young, et al., Factoring-in agglomeration of carbon nanotubes and nanofibers for better prediction of their toxicity versus asbestos, Part. FibreToxicol. 9 (2012) 10.
[181] M.E. Birch, T.A. Ruda-Eberenz, M. Chai, R. Andrews, R.L. Hatfiels, Properties that influence the specific surface areas of carbon nanotubes and nanofibers, Occup. Hyg. 57 (9) (2013) 1148–1166.
[182] G. Oberdörster, V. Castranova, B. Asgharian, P. Sayre, Inhalation exposure to carbon nanotubes (CNT) and carbon nanofibers (CNF): methodology and dosimetry, J. Toxicol. Env. Health B 18 (2015) 121–212.
[183] S. McCaldin, M. Bououdina, D.M. Grant, G.S. Walker, The effect of processing conditions on carbon nanostructures formed on an iron-based catalyst, Carbon 44 (11) (2006) 2273–2280.
[184] A. Magrez, J.W. Seo, R. Smajda, M. Mionic, L. Forro, Catalytic CVD synthesis of carbon nanotubes: towards high yield and low temperature growth, Materials 3 (11) (2010) 4871–4891.

[185] S. Fan, L. Liu, M. Liu, Monitoring the growth of carbon nanotubes by carbon isotope labelling, Nanotechnology 14 (10) (2003) 1118–1123.
[186] A.C. Ferrari, J.C. Meyer, V. Scardaci, C. Casiraghi, M. Lazzeri, F. Mauri, et al., Raman spectrum of graphene and graphene layers, Phys. Rev. Lett. (97) (2006) 187401.
[187] P.R. Kidambi, B.C. Bayer, R. Blume, Z.J. Wang, C. Baehtz, R.S. Weatherup, et al., Observing graphene grow: catalyst–graphene interactions during scalable graphene growth on polycrystalline copper, Nano Lett. 13 (10) (2013) 4769–4778.
[188] S. Gilje, S. Han, M. Wang, K.L. Wang, R.B. Kaner, A chemical route to graphene for device applications, Nano Lett. 7 (11) (2007) 3394–3398.
[189] L.C. Isett, J.M. Blakely, Segregation isosteres for carbon at the (100) surface of nickel, Surf. Sci. 58 (1976) 397.
[190] X. Li, W. Cai, J. An, S. Kim, J. Nah, D. Yang, Large-area synthesis of high-quality and uniform graphene films on copper foils, Science 324 (2009) 5932.
[191] D. Wei, Y. Liu, Y. Wang, H. Zhang, L. Huang, G. Yu, Synthesis of N-doped graphene by chemical vapor deposition and its electrical properties, Nano Lett. 9 (2009) 1752.
[192] A. Reina, X. Jia, J. Ho, D. Nezich, H. Son, V. Bulovic, et al., Layer area, few-layer graphene films on arbitrary substrates by chemical vapor deposition, Nano Lett. 9 (8) (2009) 3087.
[193] U. Khan, A. O'Neill, M. Lotya, De Sukanta, J.N. Coleman, High concentration solvent exfoliation of graphene, Small 6 (7) (2010) 864–871.
[194] Y. Si, E.T. Samulski, Synthesis of water soluble graphene, Nano Lett. 8 (6) (2008) 1679–1682.
[195] O.C. Compton, Z. An, K.W. Putz, B.J. Hong, B.G. Hauser, L.C. Brinson, et al., Additive-free hydrogelation of graphene oxide by ultrasonication, Carbon 50 (10) (2012) 3399–3406.
[196] L. Li, X. Zheng, J. Wang, Q. Sun, X. Qun, Solvent-exfoliated and functionalized graphene with assistance of supercritical carbon dioxide, ACS Sustain. Chem. Eng 1 (1) (2013) 144–151.
[197] R.A. Al-Jaboori, T. Yusaf, L. Bowtell, Energy conversion efficiency of pulsed ultrasound, Energy Proc. 75 (2015) 1560–1568.
[198] M. Zhou, T. Tian, X. Li, X. Sun, J. Zhang, P. Cui, et al., Production of graphene by liquid-phase exfoliation of intercalated graphite, Int. J. Electrochem. Sci. 9 (2014) 810–820.
[199] J.T. Han, J.I. Jang, H. Kim, J.Y. Hwang, H.K. Yoo, J.S. Woo, et al., Extremely efficient liquid exfoliation and dispersion of layered materials by unusual acoustic cavitation, Sci. Rep. 4 (2014). Article number 5133.
[200] Y. Wang, X. Tong, X. Guo, Y. Wang, G. Jin, X. Guo, Large-scale production of highly qualified graphene by ultrasonic exfoliation of expanded graphite under the promotion of $NH_4/2CO_3$ decomposition, Nanotechnology. 24 (2013).
[201] R. Durgea, R.V. Kshisagara, P. Tambe, Effect of sonication energy on the yield of graphene nanosheets by liquid-phase exfoliation of graphite, Proc. Eng. 97 (2014) 1457–1465.
[202] S. Yu, J. Liu, W. Zhu, Z.T. Hu, T.T. Lim, X. Yan, Facile room-temperature synthesis of carboxylated graphene oxide-copper sulfide nanocomposite with high photodegradation and disinfection activities under solar light irradiation, Sci. Rep. 5 (2015). Article number 16369.
[203] B. Zawisza, A. Baranik, E. Malicka, E. Talik, R. Sitco, Preconcentration of Fe(III), Co(II), Ni(II), Cu(II), Zn(II) and Pb(II) with ethylenediamine-modified graphene oxide, Microchim. Acta 183 (2016) 231–240.
[204] J. Li, J.L. Matthew, J. Vergne, E.D. Mowles, W.H. Zhong, D.M. Hercules, et al., Surface functionalization and characterization of graphitic carbon nanofibers (GCNFs), Carbon 43 (2005) 2883–2893.

[205] J. Liu, Y. Xue, L. Dai, Sulfated graphene oxide as a hole-extraction layer in high-performance polymer solar cells, J. Phys. Chem. Lett. 3 (14) (2012) 1928–1933.
[206] S. Neelakandan, N.K. Jacob, P. Kanagaraj, R.M. Sabarathinam, A. Muthumeenal, Effect of sulfonated graphene oxide on the performance enhancement of acid–base composite membranes for direct methanol fuel cells, RSC Adv. 6 (2016) 51599–51608.
[207] E. Coskun, E.A.Z. Contreras, J. Salavagione, Synthesis of sulfonated graphene/polyaniline composites with improved electroactivity, Carbon 50 (6) (2012) 2235–2243.
[208] S.V. Polschikov, P.M. Nedorezova, A.N. Klyamkina, A.A. Kovalchuk, A.M. Aladyshev, A.N. Shchegolikhin, et al., Composite materials of graphene nanoplatelets and polypropylene, prepared by in situ polymerization, J. Appl. Polym. Sci. 127 (2) (2013) 904–911.
[209] F.D. Fim, J.M. Guterres, N.R.S. Basso, G.B. Galland, Polyethylene/graphite nanocomposites obtained by in situ polymerization, J. Polym. Sci. A: Poly. Chem 48 (2010) 692–698.
[210] Y. Huang, Y. Qin, Y. Zhou, H. Niu, Z. Yu, J.Y. Dong, Polypropylene/graphene oxide nanocomposites prepared by in situ Ziegler–Natta polymerization, Chem. Mater. 22 (13) (2010) 4096–4102.
[211] E. Bekyarova, M.E. Itkis, P. Ramesh, C. Berger, M. Sprinkle, W.A. de Heer, et al., Chemical modification of epitaxial graphene: spontaneous grafting of aryl groups, J. Am. Chem. Soc. 131 (4) (2009) 1336–1337.
[212] J.R. Lomeda, C.D. Doyle, D.V. Kosynkin, W.F. Hwang, J.M. Tour, Diazonium functionalization of surfactant-wrapped chemically converted graphene sheets, J. Am. Chem. Soc. 130 (48) (2008) 16201–16206.
[213] J.P. Tessonnier, M.A. Barteau, Dispersion of alkyl-chain-functionalized reduced graphene oxide sheets in nonpolar solvents, Langmuir. 28 (16) (2012) 6691–6697.
[214] F. Li, Y. Bao, J. Chai, Q. Zhang, D. Han, L. Niu, Synthesis and application of widely soluble graphene sheets, Langmuir. 26 (14) (2010) 12314–12320.
[215] B. Zhang, W. Ning, J. Zhang, J. Qiao, J. He, C.Y. Liu, Stable dispersions of reduced graphene oxide in ionic liquids, J. Mater. Chem. 26 (2010) 5401–5403.
[216] Y. Xu, Z. Liu, X. Zhang, Y. Wang, J. Tian, Y. Huang, et al., Graphene hybrid material covalently functionalized with porphyrin: synthesis and optical limiting property, Adv. Mater. 21 (2009) 1275–1279.
[217] Z.B. Liu, Y.F. Xu, X.Y. Zhang, Y.S. Zhang, J.G. Tian, Porphyrin and fullerene covalently functionalized graphene hybrid materials with large nonlinear optical properties, J. Phys. Chem. B. 113 (29) (2009) 9681–9686.
[218] X. Zhang, Y. Feng, D. Huang, Y. Li, W. Feng, Investigation of optical modulated conductance effects based on a graphene oxide–azobenzene hybrid, Carbon 48 (11) (2010) 3226–3241.
[219] N. Karousis, A.S.D. Sandanayaka, T. Hasobe, S.P. Economopoulos, Graphene oxide with covalently linked porphyrin antennae: synthesis, characterization and photophysical properties, J. Mater. Chem. 21 (2011) 109–117.
[220] W. Li, X.Z. Tang, H.B. Zhang, Z.G. Jiang, Z.Z. Yu, X.S. Du, Simultaneous surface functionalization and reduction of graphene oxide with octadecylamine for electrically conductive polystyrene composites, Carbon 49 (14) (2011) 4724–4730.
[221] X. Wang, Y. Hu, L. Song, H. Yang, W. Xing, H. Lu, In situ polymerization of graphene nanosheets and polyurethane with enhanced mechanical and thermal properties, J. Mater. Chem. 21 (2011) 4222–4227.
[222] K.P. Pramoda, H. Hussain, H.M. Koh, H.R. Tan, C.B. He, Covalent bonded polymer-graphene nanocomposites, J. Polym. Sci. A: Polym. Chem. 48 (2010) 4262–4267.

[223] N.D. Luong, U. Hippi, J.T. Korhonen, A.J. Soininen, J. Ruokolainen, L.S. Johansson, et al., Enhanced mechanical and electrical properties of polyimide film by graphene sheets via in situ polymerization, Polymer 52 (23) (2011) 5237–5242.
[224] L.M. Veca, F. Lu, M.J. Meziani, L. Cao, P. Zhang, G. Qi, et al., Polymer functionalization and solubilization of carbon nanosheets, Chem. Commun (18) (2009) 2565–2567.
[225] Q. Yang, X. Pan, K. Clarke, K. Li, Covalent functionalization of graphene with polysaccharides, Ind. Eng. Chem. Res. 51 (2012) 310–317.
[226] N. Hu, L. Meng, R. Gao, Y. Wang, J. Chai, Z. Yang, et al., A facile route for the large scale fabrication of graphene oxide papers and their mechanical enhancement by cross-linking with glutaraldehyde, Nano-Micro Lett. 3 (4) (2011) 215–222.
[227] H.J. Salavagione, M.A. Gomez, G. Martinez, Polymeric modification of graphene through esterification of graphite oxide and poly(vinyl alcohol), Macromolecules. 42 (2009) 6331–6334.
[228] H.J. Salavagione, G. Martinez, Importance of covalent linkages in the preparation of effective reduced graphene oxide–poly(vinyl chloride) nanocomposites, Macromolecules. 44 (8) (2011) 2685–2692.
[229] Y. Pan, N.G. Sahoo, L. Li, The application of graphene oxide in drug delivery, Adv. Funct. Mater. 21 (2011) 2754–2763.
[230] Z. Jin, T.P. McNicholas, C.J. Shih, Q.H. Wang, G.L.C. Paulus, A.J. Hilmer, et al., Click chemistry on solution-dispersed graphene and monolayer CVD graphene, Chem. Mater. 23 (2011) 3362–3370.
[231] S. Sun, Y. Cao, J. Feng, P. Wu, Click chemistry as a route for the immobilization of well-defined polystyrene onto graphene sheets, J. Mater. Chem. 20 (2010) 5605–5607.
[232] Y.S. Ye, Y.N. Chen, J.S. Wang, J. Rick, Y.J. Huang, F.C. Chang, et al., Versatile grafting approaches to functionalizing individually dispersed graphene nanosheets using RAFT polymerization and click chemistry, Chem. Mater. 24 (2012) 2987–2997.
[233] L. Kou, H. He, C. Gao, Click chemistry approach to functionalize two-dimensional macromolecules of graphene oxide nanosheets, Nano-Micro Lett. 2 (3) (2010) 177–183.
[234] Z. Xu, C. Gao, In situ polymerization approach to graphene-reinforced nylon-6 composites, Macromolecules. 43 (2010) 67.
[235] S. Sun, P. Wu, A one step strategy for thermal and pH responsive graphene oxide interpenetrating polymer hydrogel networks, J. Mater. Chem. 21 (2011) 4095–4097.
[236] J. Jangbarwala, Composite porous dehumidifying material for an HVAC, US Patent 9,211,499, 2015.
[237] Y.X. Xu, H. Bai, G. Lu, C. Li, G. Shi, Flexible graphene films via the filtration of water-soluble noncovalent functionalized graphene sheets, J. Am. Chem. Soc. 130 (2008) 5856.
[238] Q.H. Wang, M.C. Hersam, Room-temperature molecular-resolution characterization of self-assembled organic monolayers on epitaxial graphene, Nat. Chem. 1 (2009) 206–211.
[239] Q. Su, S. Pang, V. Alijani, C. Li, X. Feng, K. Mullen, Composites of graphene with large aromatic molecules, Adv. Mater. 21 (2009) 3191–3195.
[240] J. Liang, Y. Huang, L. Zhang, Y. Wang, Y. Ma, T. Guo, et al., Molecular-level dispersion of graphene into poly(vinyl alcohol) and effective reinforcement of their nanocomposites, Adv. Funct. Mater. 19 (14) (2009) 2297–2302.
[241] R.K. Layek, S. Samanta, A.K. Nandi, The physical properties of sulfonated graphene/poly(vinyl alcohol) composites, Carbon 50 (3) (2012) 815–827.

[242] J. Liu, W. Yang, L. Tao, D. Li, C. Boyer, T. Davis, Thermosensitive graphene nanocomposites formed using pyrene-terminal polymers made by RAFT polymerization, J. Polym. Sci. A Polym. Chem. 48 (2010) 425–433.

[243] J. Liu, L. Tao, W. Yang, D. Li, C. Boyer, R. Wuhrer, et al., Synthesis, characterization, and multilayer assembly of pH sensitive graphene-polymer nanocomposites, Langmuir 26 (2010) 10068–10075.

[244] W. Zoujun, R. Pan, Y. Hou, Y. Yang, Y. Liu, Graphene-supported Pd catalyst for highly selective hydrogenation of resorcinol to 1, 3-cyclohexanedione through giant π-conjugate interactions, Sci. Rep. 5 (2015) 15664. ISSN 2045-2322.

[245] R. Krishnamoorti, R.A. Vaia (Eds.), Polymer nano composites, synthesis, characterization and modeling, in: ACS Symposium Series, vol. 804, 2001.

[246] M.R. Piggott, Load Bearing Fibre Comp, Pergamon Press, Oxford, 1980.

[247] R.A. Vaia, Polymer nanocomposites open a new dimension for plastics and composites, AMTIAC Newslett. 6 (2002) 17–24.

[248] P. Potschle, T.D. Fornes, D.R. Paul, Rheological behavior of multiwalled carbon nanotube/polycarbonate composites, Polymer 43 (11) (2002) 3247–3255.

[249] E. Hammel, X. Tang, M. Trampert, T. Schmitt, K. Mauthner, A. Eder, et al., Carbon nanofibers for composite applications, Carbon 42 (5–6) (2004) 1153–1158.

[250] S.M. Rhodes, B. Higgins, Y. Xu, J. Brittain, Hyperbranched polyol/carbon nanofiber composites, Polymer 48 (6) (2007) 1500–1509.

[251] P.C. Painter, M.M. Coleman, et al., Weak Specific Interractions, in: D.R. Paul, C. B. Bucknall (Eds.), Polymer Blends, Vol. 1, Wiley, New York, NY, 2000, p. 93.

[252] M.M. Coleman, J.F. Graf, P.C. Painter, Specific Interactions and the Miscibility of Polymer Blends, Technomic Publishing, Lancaster, PA, 1991.

[253] M.M. Coleman, L.A. Narvett, P.C. Painter, A counterintuitive observation concerning hydrogen bonding in polymer blends, Polymer 39 (23) (1998) 5867–5869.

[254] J.Y. Lee, P.C. Painter, M.M. Coleman, Hydrogen bonding in polymer blends. 4. Blends involving polymers containing methacrylic acid and vinylpyridine groups, Macromolecules 21 (4) (1988) 954–960.

[255] J.A. Moore, S. Kaur, Blends of poly(amide-enaminonitrile) with poly(ethylene oxide), poly(4-vinylpyridine), and poly(*N*-vinylpyrrolidone), Macromolecules 31 (2) (1998) 328–335.

[256] Y.E. Kirsh, Water Soluble Poly(*N*-vinylamides), Wiley, New York, NY, 1998.

[257] G. Polacco, M.G. Cascone, L. Petarca, A. Peretti, Thermal behaviour of poly (methacrylic acid)/poly(*N*-vinyl-2-pyrrolidone) complexes, Eur. Polym. J. 36 (2000) 2541.

[258] N. Cassu, M.I. Felisberti, Poly(vinyl alcohol) and poly(vinyl pyrrolidone) blends: miscibility, microheterogeneity and free volume change, Polymer 38 (1997) 3907.

[259] A. Singh, M.C. Saxena, An equation to calculate the equilibrium constant of 1:1complexes from dielectric relaxation measurements, J. Mol. Liq. 25 (1983) 81.

[260] A. Singh, R. Misra, J.P. Shukla, M.C. Saxena, Evaluation of the thermodynamic parameters for association process of hydrogen bond complexes from the dielectric relaxation measurements, J. Mol. Liq. 26 (1983) 29.

[261] P. Sivagurunathan, K. Dharmalingam, K. Ramachandran, Solvent effects on hydrogen bonding between ethyl methacrylate and butanol, Phys. Chem. 219 (2005) 38.

[262] R.E. Kagarise, K.B. Whetsel, Solvent effects on infrared frequencies—III: the carbony bands of ethyl acetate, ethyl trichloroacetate and ethyl trifluoroacetate, Spectrochim. Acta 18 (1962) 341.

[263] M.G.K. Pillai, K. Ramaswamy, S.G. Gnanadesikan, Influence of solvents on infrared frequencies-the carbonyl bands of phenyl acetate, methyl ethyl ketone, and benzaldehyde, Austr. J. Chem. 3 (1966) 1089.

[264] G.G. Tibbetts, M.L. Lake, K.L. Strong, B.P. Rice, A review of the fabrication and properties of vapor-grown carbon nanofiber/polymer composites, Compos. Sci. Technol. 67 (2007) 1709–1718.
[265] D.G. Glasgow, G.G. Tibbetts, J.J. Matuszewski, K.R. Walters, M.L. Lake, Surface treatment of carbon nano fibers for improved composite mechanical properties, in: International Society for Advancement of Materials and Process Engineering (SAMPE) Symposium and Exhibition, Long Beach, CA, May 2004, pp. 16–20.
[266] P.V. Lakshminarayanan, H. Toghiani, C.U. Pittman Jr., Nitric acid oxidation of vapor grown carbon nanofibers, Carbon 42 (2004) 2422–2433.
[267] K. Lafdi, M. Matthew, Carbon nanofibers as a nano-reinforcement for polymeric nanocomposites, in: the 35th International SAMPE Technical Conference, 2003.
[268] D.G. Glasgow, M.L. Lake, Production of high surface energy, high surface area vapor grown carbon fiber, US Patent 6,506,355, 2003.
[269] S. Kumar, B. Lively, L.L. Sun, B. Li, W.H. Zhong, Highly dispersed and electrically conductive polycarbonate/oxidized carbon nanofiber composites for electrostatic dissipation applications, Carbon 48 (2010) 3846–3857.
[270] Y.K. Choi, K.I. Sugimoto, S.M. Song, M. Endo, Mechanical and thermal properties of vapor-grown carbon nanofiber and polycarbonate composite sheets, Mater. Lett. 59 (2005) 3514–3520.
[271] H. Kim, C.W. Macosko, Processing-property relationships of polycarbonate/grapheme composites, Polymer 50 (2009) 3797–3809.
[272] M. Yoonessi, E. Boyd, D.J. Quade, D. Scheiman, Graphene Polymer Nanocomposites, NASA-Ohio Aerospace Institute-The NASA Aeronautics-Subsonic Fixed Wing Program, (2011). Contract NNC07BA13B.
[273] G. Gedler, M. Antunes, V. Realinho, J.I. Velasco, Thermal stability of polycarbonate-graphene nanocomposite foams, Polym. Degrad. Stab. 97 (8) (2012) 1297–1304.
[274] M. Yoonessi, J.R. Gaier, Highly conductive multifunctional graphene polycarbonate nanocomposites, ACS Nano 4 (12) (2010) 7211–7220.
[275] J.R. Potts, S. Murali, Y. Zhu, X. Zhao, R.S. Ruoff, Microwave-exfoliated graphite oxide/polycarbonate composites, Macromolecules 44 (16) (2011) 6488–6495.
[276] R. Mahendran, D. Sridharan, K. Santhakumar, T.A. Selvakumar, P. Rajasekar, J.H. Jang, Graphene oxide reinforced polycarbonate nanocomposite films with antibacterial properties, Ind. J. Mater. Sci. 2016 (2016) 10 pages. Article ID 4169409.
[277] C. Leer, F.W.J. Van Hattum, A.G. Cunha, O.S. Carneiro, C.A. Bernardo, Tailored shear extrusion of carbon nanofibre/polyamide composites and its effect on electrical percolation threshold, Plast. Rubber Compos. 35 (6/7) (2006) 268–275.
[278] R.D. Goodrich, M.L. Shofner, R.J.M. Hague, M. McClelland, M.R. Schlea, R.B. Johnson, et al., Processing of a polyamide-12/carbon nanofibre composite by laser sintering, Polym. Test. 30 (2011) 94–100.
[279] X. Jinag, L.T. Drzal, Multifunctional high density polyethylene nanocomposites produced by incorporation of exfoliated graphite nanoplatelets 1: Morphology and mechanical properties, Polym. Compos. (2010) 1091–1098.
[280] I.C. Finnegan, G.G. Tibbetts, Electrical conductivity of vapor-grown carbon fiber/thermoplastic composites, J. Mater. Res. 16 (6) (2001) 1668–1674.
[281] G.G. Tibbetts, I.C. Finegan, Mechanical and electrical propertiesof vapor-grown carbon fiber thermoplastic composites, Mol. Cryst. Liq. Cryst. 387 (2002) 129–133.
[282] K. Enomoto, T. Yasuhara, N. Ohtake, Mechanical properties of injection-molded composites of carbon nanofibers in polypropylene matrix, New Diam. Front. Carbon Technol. 15 (2005) 2.

[283] M. Chipara, K. Lozano, A. Hernandez, M. Chipara, TGA analysis of polypropylene carbon nanofibers composites, Polym. Degrad. Stab. 93 (2008) 871–876.
[284] M.S. Dresselhaus, Graphite Fibers and Filaments, *Springer*, Berlin, 1988, p. 109.
[285] Y.M. Chen, J.M. Ting, Ultra high thermal conductivity polymer composites, Carbon 40 (3) (2002) 359–362.
[286] <http://www.nanosperse.com>
[287] <http://www.ctd-materials.com>
[288] <http://www.electrovac.com>
[289] J. Vetter, P. Novák, M.R. Wagner, C. Veit, K.C. Möller, J.O. Besenhard, et al., Ageing mechanisms in lithium-ion batteries, J. Power Sources 147 (1) (2005) 269–281.
[290] J.R. Dahn, Lithium Batteries. New Materials, Developments, and Perspectives, Elsevier, Amsterdam, 1994, p. 1.
[291] M.C. Smart, B.V. Ratnakumar, S. Surampudi, Electrolytes for low-temperature lithium batteries based on ternary mixtures of aliphatic carbonates, J. Electrochem. Soc. 146 (1999) 486.
[292] E.J. Plichita, W.K. Behl, A low-temperature electrolyte for lithium and lithium-ion batteries, J. Power Sources 88 (2000) 192.
[293] B.V. Ratnakumar, M.C. Smart, S. Surampudi, Effects of SEI on the kinetics of lithium intercalation, J. Power Sources 97–98 (2001) 137.
[294] S.S. Zhang, K. Xu, T.R. Jow, Low temperature performance of graphite electrode in Li-ion cells, Electrochim. Acta. 48 (2002) 241.
[295] H. Lin, D. Chua, M. Salomon, H.C. Shiao, M. Hendrickson, E. Plichta, et al., Low-temperature behavior of Li-Ion Cells, Electrochem. Solid State Lett. 4 (2001) A71.
[296] C. Wang, A.J. Appleby, F.E. Little, Low-temperature characterization of lithium-ion carbon anodes via microperturbation measurement, J. Electrochem. Soc. 149 (2002) A754.
[297] L. Zhao, I. Watanabe, T. Doi, S. Okada, J.I. Yamaki, TG-MS analysis of solid electrolyte interphase (SEI) on graphite negative-electrode in lithium-ion batteries, J. Power Sources 161 (2006) 1275–1280.
[298] K. Mariguchi, Y. Itoh, S. Munetoh, K. Kamel, M. Abe, A. Omaru, et al., Nanotube-like surface structure in graphite anodes for lithium-ion secondary batteries, Phys. B 323 (2002) 127–129.
[299] S.H. Yoon, C.W. Park, H. Yang, Y. Korai, I. Mochida, R.T.K. Baker, et al., Novel carbon nanofibers of high graphitization as anodic materials for lithium ion secondary batteries, Carbon 42 (1) (2004) 21–32.
[300] E. Peled, D.B. Tow, A. Merson, A. Gladkich, L. Burstein, D. Golodnitsky, Composition, depth profiles and lateral distribution of materials in the SEI built on HOPG-TOF SIMS and XPS studies, J. Power Sources 97-98 (2001) 52–57.
[301] E. Peled, et al., Handbook of Battery Materials, Wiley–VCH, NY, 1999, p. 419.
[302] S. Sato, R. Takahashi, T. Sodesawa, F. Nozaki, X.Z. Jin, S. Suzuki, et al., Mass-transfer limitation in mesopores of Ni–MgO catalyst in liquid-phase hydrogenation, J. Catal. 191 (2) (2000) 261–270.
[303] P. Serp, M. Corrias, P. Kalck, Carbon nanotubes and nanofibers in catalysis, Appl. Catal., A 253 (2) (2003) 337–358.
[304] M. Menon, A.N. Andriotis, G.E. Froudakis, Curvature dependence of the metal catalyst atom interaction with carbon nanotubes walls, Chem. Phys. Lett. 320 (5) (2000) 425–434.
[305] M. Ouyang, J.L. Huang, C.M. Lieber, Fundamental electronic properties and applications of single-walled carbon nanotubes, Acc. Chem. Res. 35 (12) (2002) 1018–1025.

[306] F. Nunzi, F. Mercuri, A. Sgamellotti, N. Re, The coordination chemistry of carbon nanotubes: a density functional study through a cluster model approach, J. Phys. Chem. B. 106 (41) (2002) 10622–10633.
[307] R. Vieira, Carbon nanofibers as macro-structured catalytic support, in: A. Kumar (Ed.), Nanofibers, InTech, (2010) http://dx.doi.org/10.5772/8145. Available from: http://www.intechopen.com/books/nanofibers/carbon-nanofibers-as-macro-structured-catalytic-support.
[308] L.I.U. Pingle, L. Lefferts, Preparation of carbon nano-fiber washcoat on porous silica foam as structured catalyst support, Chin. J. Chem. Eng. 14 (3) (2006) 294–300.
[309] X. Xu, N.M. Rodriguez, R.T.K. Baker, Ethylene oxidation over Ag supported on novel carbon nano-structured supports, React. Kinet. Catal. Lett. 87 (2) (2006) 305–312.
[310] T.C. Rocha, A. Oestereich, D.V. Demidov, M. Hävecker, S. Zafeiratos, G. Weinberg, et al., The silver–oxygen system in catalysis: new insights by near ambient pressure X-ray photoelectron spectroscopy, Phys. Chem. Chem. Phys. 14 (13) (2012) 4554–4564.
[311] P.J. Van den Hoek, E.J. Baerends, R.A. Van Santen, Ethylene epoxidation on silver (110): the role of subsurface oxygen, J. Phys. Chem. 93 (17) (1989) 6469–6475.
[312] I.A.W. Filot, R.A. van Santen, E.J.M. Hensen, Quantum chemistry of the Fischer–Tropsch reaction catalysed by a stepped ruthenium surface, Catal. Sci. Technol. 4 (9) (2014) 3129–3140.
[313] C.J. Weststrate, J. van de Loosdrecht, J.W. Niemantsverdriet, Spectroscopic insights into cobalt-catalyzed Fischer-Tropsch synthesis: a review of the carbon monoxide interaction with single crystalline surfaces of cobalt, J. Catal. 342 (2016) 1–16.
[314] G.P. Van Der Laan, A.A.C.M. Beenackers, Kinetics and selectivity of the Fischer–Tropsch synthesis: a literature review, Catal. Rev. 41 (3-4) (1999) 255–318.
[315] R.C. Brady III, R. Pettit, Mechanism of the Fischer-Tropsch reaction. The chain propagation step, J. Am. Chem. Soc. 103 (5) (1981) 1287–1289.
[316] G. Henrici-Olivé, S. Olive, The Fischer-Tropsch synthesis: molecular weight distribution of primary products and reaction mechanism, Angew. Chem. Int. Ed. Engl. 15 (3) (1976) 136–141.
[317] P. Biloen, W.M.H. Sachtler, Mechanism of hydrocarbon synthesis over Fischer-Tropsch catalysts, Adv. Catal. 30 (1981) 165–216.
[318] M.E. Dry, Practical and theoretical aspects of the catalytic Fischer-Tropsch process, Appl. Catal. A 138 (2) (1996) 319–344.
[319] B.H. Davis, Fischer–Tropsch synthesis: current mechanism and futuristic needs, Fuel Process. Technol. 71 (1) (2001) 157–166.
[320] O.R. Inderwildi, S.J. Jenkins, D.A. King, Fischer-Tropsch mechanism revisited: alternative pathways for the production of higher hydrocarbons from synthesis gas, J. Phys. Chem. C 112 (5) (2008) 1305–1307.
[321] M. Ojeda, R. Nabar, A.U. Nilekar, A. Ishikawa, M. Mavrikakis, E. Iglesia, CO activation pathways and the mechanism of Fischer–Tropsch synthesis, J. Catal. 272 (2) (2010) 287–297.
[322] J. Patzlaff, Y. Liu, C. Graffmann, J. Gaube, Studies on product distributions of iron and cobalt catalyzed Fischer–Tropsch synthesis, Appl. Catal. A 186 (1) (1999) 109–119.
[323] R.B. Anderson, R.A. Friedel, H.H. Storch, Fischer-Tropsch reaction mechanism involving stepwise growth of carbon chain, J. Chem. Phys. 19 (3) (1951) 313–319.
[324] V. Ponec, Some aspects of the mechanism of methanation and Fischer-Tropsch synthesis, Catal. Rev. Sci. Eng. 18 (1) (1978) 151–171.

[325] B.H. Davis, Fischer–Tropsch synthesis: reaction mechanisms for iron catalysts, Catal. Today 141 (1) (2009) 25–33.
[326] S. Storsæter, D. Chen, A. Holmen, Microkinetic modelling of the formation of C 1 and C 2 products in the Fischer–Tropsch synthesis over cobalt catalysts, Surf. Sci. 600 (10) (2006) 2051–2063.
[327] H. Schulz, E. vein Steen, M. Claeys, Selectivity and mechanism of Fischer-Tropsch synthesis with iron and cobalt catalysts, Stud. Surf. Sci. Catal. 81 (1994) 455–460.
[328] Ø. Borg, S. Eri, E.A. Blekkan, S. Storsæter, H. Wigum, E. Rytter, et al., Fischer–Tropsch synthesis over γ-alumina-supported cobalt catalysts: effect of support variables, J. Catal. 248 (1) (2007) 89–100.
[329] W. Chu, P.A. Chernavskii, L. Gengembre, G.A. Pankina, P. Fongarland, A.Y. Khodakov, Cobalt species in promoted cobalt alumina-supported Fischer–Tropsch catalysts, J. Catal. 252 (2) (2007) 215–230.
[330] A.M. Hilmen, D. Schanke, K.F. Hanssen, A. Holmen, Study of the effect of water on alumina supported cobalt Fischer–Tropsch catalysts, Appl. Catal. A 186 (1) (1999) 169–188.
[331] J.S. Girardon, E. Quinet, A. Griboval-Constant, P.A. Chernavskii, L. Gengembre, A.Y. Khodakov, Cobalt dispersion, reducibility, and surface sites in promoted silica-supported Fischer–Tropsch catalysts, J. Catal. 248 (2) (2007) 143–157.
[332] J. Van de Loosdrecht, M. Van der Haar, A.M. Van der Kraan, A.J. Van Dillen, J.W. Geus, Preparation and properties of supported cobalt catalysts for Fischer-Tropsch synthesis, Appl. Catal. A 150 (2) (1997) 365–376.
[333] C. Lancelot, V.V. Ordomsky, O. Stéphan, M. Sadeqzadeh, H. Karaca, M. Lacroix, et al., Direct evidence of surface oxidation of cobalt nanoparticles in alumina-supported catalysts for Fischer–Tropsch synthesis, ACS Catal. 4 (12) (2014) 4510–4515.
[334] L. Ji, J. Lin, H.C. Zeng, Metal-support interactions in Co/Al_2O_3 catalysts: a comparative study on reactivity of support, J. Phys. Chem. B. 104 (8) (2000) 1783–1790.
[335] G.L. Bezemer, J.H. Bitter, H.P. Kuipers, H. Oosterbeek, J.E. Holewijn, X. Xu, et al., Cobalt particle size effects in the Fischer-Tropsch reaction studied with carbon nanofiber supported catalysts, J. Am. Chem. Soc. 128 (12) (2006) 3956–3964.
[336] J.M.G. Carballo, J. Yang, A. Holmen, S. García-Rodríguez, S. Rojas, M. Ojeda, et al., Catalytic effects of ruthenium particle size on the Fischer–Tropsch Synthesis, J. Catal. 284 (1) (2011) 102–108.
[337] R.A. van Santen, A.J. Markvoort, Catalyst nano-particle size dependence of the Fischer–Tropsch reaction, Faraday. Discuss. 162 (2013) 267–279.
[338] M.E. Dry, High quality diesel via the Fischer–Tropsch process—a review, Journal of Chemical Technology and Biotechnology 77 (1) (2002) 43–50.
[339] P.J. Van Berge, J. Van de Loosdrecht, S. Barradas, A.M. Van Der Kraan, Oxidation of cobalt based Fischer–Tropsch catalysts as a deactivation mechanism, Catal. Today 58 (4) (2000) 321.
[340] N.E. Tsakoumis, M. Rønning, Ø. Borg, E. Rytter, A. Holmen, Deactivation of cobalt based Fischer–Tropsch catalysts: a review, Catal. Today 154 (3) (2010) 162–182.
[341] D.J. Moodley, J. Van de Loosdrecht, A.M. Saib, M.J. Overett, A.K. Datye, J.W. Niementsverdriet, Carbon deposition as a deactivation mechanism of cobalt-based Fischer–Tropsch synthesis catalysts under realistic conditions, Appl. Catal. A 354 (1) (2009) 102–110.

[342] R. Philippe, M. Lacroix, L. Dreibine, C. Pham-Huu, D. Edouard, S. Savin, et al., Effect of structure and thermal properties of a Fischer–Tropsch catalyst in a fixed bed, Catal. Today 147 (2009) S305–S312.
[343] Y. Liu, C. Pham-Huu, P. Nguyen, C. Pham, Catalyst supports made from silicon carbide covered with TiO_2 for Fischer-Tropsch synthesis, US Patent Application No. 14/411,543, 2013.
[344] Z. Yu, Ø. Borg, D. Chen, B.C. Enger, V. Frøseth, E. Rytter, et al., Carbon nanofiber supported cobalt catalysts for Fischer–Tropsch synthesis with high activity and selectivity, Catal. Lett. 109 (1-2) (2006) 43–47.
[345] S. Zarubova, S. Rane, J. Yang, Y. Yu, Y. Zhu, D. Chen, et al., Fischer–Tropsch synthesis on hierarchically structured cobalt nanoparticle/carbon nanofiber/carbon felt composites, ChemSusChem 4 (7) (2011) 935–942.
[346] G.L. Bezemer, P.B. Radstake, V. Koot, A.J. Van Dillen, J.W. Geus, K.P. De Jong, Preparation of Fischer–Tropsch cobalt catalysts supported on carbon nanofibers and silica using homogeneous deposition-precipitation, J. Catal. 237 (2) (2006) 291–302.
[347] G.L. Bezemer, A. Van Laak, A.J. Van Dillen, K.P. De Jong, Cobalt supported on carbon nanofibers-a promising novel Fischer-Tropsch catalyst, Stud. Surf. Sci. Catal. 147 (2004) 259–264.
[348] M. Trépanier, A. Tavasoli, A.K. Dalai, N. Abatzoglou, Fischer–Tropsch synthesis over carbon nanotubes supported cobalt catalysts in a fixed bed reactor: influence of acid treatment, Fuel Process. Technol. 90 (3) (2009) 367–374.
[349] A.Y. Khodakov, Fischer-Tropsch synthesis: relations between structure of cobalt catalysts and their catalytic performance, Catal. Today 144 (3) (2009) 251–257.
[350] A.H.M.A.D. Tavasoli, K. Sadagiani, F. Khorashe, A.A. Seifkordi, A.A. Rohani, A. Nakhaeipour, Cobalt supported on carbon nanotubes—a promising novel Fischer–Tropsch synthesis catalyst, Fuel Process. Technol. 89 (5) (2008) 491–498.
[351] C. Park, R.T.K. Baker, Catalytic behavior of graphite nanofiber supported nickel particles. 2. The influence of the nanofiber structure, J. Phys. Chem. B. 102 (26) (1998) 5168–5177.
[352] T.G. Ros, A.J. Van Dillen, J.W. Geus, D.C. Koningsberger, Modification of carbon nanofibres for the immobilization of metal complexes: a case study with rhodium and anthranilic acid, Chem. Eur. J. 8 (13) (2002) 2868–2878.
[353] M.J. Ledoux, R. Vieira, C. Pham-Huu, N. Keller, New catalytic phenomena on nanostructured (fibers and tubes) catalysts, J. Catal. 216 (1) (2003) 333–342.
[354] F. Salman, C. Park, R.T.K. Baker, Hydrogenation of crotonaldehyde over graphite nanofiber supported nickel, Catal. Today 53 (3) (1999) 385–394.
[355] J.M. Nhut, R. Vieira, L. Pesant, J.P. Tessonnier, N. Keller, G. Ehret, et al., Synthesis and catalytic uses of carbon and silicon carbide nanostructures, Catal. Today 76 (1) (2002) 11–32.
[356] P. Liu, H. Xie, S. Tan, K. You, N. Wang, H.A. Luo, Carbon nanofibers supported nickel catalyst for liquid phase hydrogenation of benzene with high activity and selectivity, React. Kinet. Catal. Lett. 97 (1) (2009) 101–108.
[357] P. Tribolet, L. Kiwi-Minsker, Palladium on carbon nanofibers grown on metallic filters as novel structured catalyst, Catal. Today 105 (3) (2005) 337–343.
[358] A.S. Nagpure, L. Gurrala, P. Gogoi, S.V. Chilukuri, Hydrogenation of cinnamaldehyde to hydrocinnamaldehyde over Pd nanoparticles deposited on nitrogen-doped mesoporous carbon, RSC Adv. 6 (50) (2016) 44333–44340.
[359] L. Truong-Phuoc, T. Truong-Huu, L. Nguyen-Dinh, W. Baaziz, T. Romero, D. Edouard, et al., Silicon carbide foam decorated with carbon nanofibers as catalytic stirrer in liquid-phase hydrogenation reactions, Appl. Catal. A 469 (2014) 81–88.

[360] A.M. Zhang, J.L. Dong, Q.H. Xu, H.K. Rhee, X.L. Li, Palladium cluster filled in inner of carbon nanotubes and their catalytic properties in liquid phase benzene hydrogenation, Catal. Today 93 (2004) 347–352.
[361] T.P. Vispute, H. Zhang, A. Sanna, R. Xiao, G.W. Huber, Renewable chemical commodity feedstocks from integrated catalytic processing of pyrolysis oils, Science 330 (6008) (2010) 1222–1227.
[362] E.D. Christensen, G.M. Chupka, J. Luecke, T. Smurthwaite, T.L. Alleman, K. Iisa, et al., Analysis of oxygenated compounds in hydrotreated biomass fast pyrolysis oil distillate fractions, Energy Fuels 25 (11) (2011) 5462–5471.
[363] W.J. Liu, X.S. Zhang, Y.C. Qv, H. Jiang, H.Q. Yu, Bio-oil upgrading at ambient pressure and temperature using zero valent metals, Green Chem. 14 (8) (2012) 2226–2233.
[364] M.I. Jahirul, M.G. Rasul, A.A. Chowdhury, N. Ashwath, Biofuels production through biomass pyrolysis—a technological review, Energies 5 (12) (2012) 4952–5001.
[365] J. Wildschut, J. Arentz, C.B. Rasrendra, R.H. Venderbosch, H.J. Heeres, Catalytic hydrotreatment of fast pyrolysis oil: model studies on reaction pathways for the carbohydrate fraction, Environ. Progr. Sustain. Energy 28 (3) (2009) 450–460.
[366] L. Braeckman, W. Prins, J. Pieters. Catalytic upgrading of biomass fast pyrolysis oils, in: 14th Symposium on Applied Biological Sciences, Vol. 73, No. 1, 2008, pp. 3–7.
[367] T.J. Benson, P.R. Daggolu, R.A. Hernandez, S. Liu, M.G. White, Catalytic deoxygenation chemistry: upgrading of liquids derived from biomass processing, Adv. Catal. 56 (56) (2013) 187–353.
[368] J. Wildschut, I. Melian-Cabrera, H.J. Heeres, Catalyst studies on the hydrotreatment of fast pyrolysis oil, Appl. Catal. B Environ. 99 (1) (2010) 298–306.
[369] Z. Su-Ping, Study of hydrodeoxygenation of bio-oil from the fast pyrolysis of biomass, Energy Sources 25 (1) (2003) 57–65.
[370] A.V. Bridgwater, Review of fast pyrolysis of biomass and product upgrading, Biomass Bioenergy 38 (2012) 68–94.
[371] D.C. Elliott, T.R. Hart, G.G. Neuenschwander, L.J. Rotness, A.H. Zacher, Catalytic hydroprocessing of biomass fast pyrolysis bio-oil to produce hydrocarbon products, Environ. Progr. Sustain. Energy 28 (3) (2009) 441–449.
[372] Y. Elkasabi, C.A. Mullen, A.L. Pighinelli, A.A. Boateng, Hydrodeoxygenation of fast-pyrolysis bio-oils from various feedstocks using carbon-supported catalysts, Fuel Process. Technol. 123 (2014) 11–18.
[373] Q. Zhang, J. Chang, T. Wang, Y. Xu, Review of biomass pyrolysis oil properties and upgrading research, Energy Convers. Manage. 48 (1) (2007) 87–92.
[374] S.B. Jones, C. Valkenburg, C.W. Walton, D.C. Elliott, J.E. Holladay, D.J. Stevens, et al., Production of gasoline and diesel from biomass via fast pyrolysis, hydrotreating and hydrocracking: a design case, Pacific Northwest National Laboratory, Richland, 2009, pp. 1–76.
[375] A.W. Scaroni, R.G. Jenkins, P.L. Walker, Coke deposition on Co-Mo/Al_2O_3 and Co-Mo/C catalysts, Appl. Catal. 14 (1985) 173–183.
[376] D. Kubička, J. Horáček, Deactivation of HDS catalysts in deoxygenation of vegetable oils, Appl. Catal. A 394 (1) (2011) 9–17.
[377] F. Diez, B.C. Gates, J.T. Miller, D.J. Sajkowski, S.G. Kukes, Deactivation of a nickel-molybdenum/gamma.-alumina catalyst: influence of coke on the hydroprocessing activity, Ind. Eng. Chem. Res. 29 (10) (1990) 1999–2004.
[378] C. Wang, J. Qiu, C. Liang, L. Xing, X. Yang, Carbon nanofiber supported Ni catalysts for the hydrogenation of chloronitrobenzenes, Catal. Commun. 9 (8) (2008) 1749–1753.

[379] J. Han, H. Sun, J. Duan, Y. Ding, H. Lou, X. Zheng, Palladium-catalyzed transformation of renewable oils into diesel components, Adv. Synth. Catal. 352 (11–12) (2010) 1805–1809.
[380] M. Chiappero, P.T.M. Do, S. Crossley, L.L. Lobban, D.E. Resasco, Direct conversion of triglycerides to olefins and paraffins over noble metal supported catalysts, Fuel 90 (3) (2011) 1155–1165.
[381] T. Morgan, D. Grubb, E. Santillan-Jimenez, M. Crocker, Conversion of triglycerides to hydrocarbons over supported metal catalysts, Topics Catal. 53 (11–12) (2010) 820–829.
[382] I. Kubičková, M. Snåre, K. Eränen, P. Mäki-Arvela, D.Y. Murzin, Hydrocarbons for diesel fuel via decarboxylation of vegetable oils, Catal. Today 106 (1) (2005) 197–200.
[383] G.N. da Rocha Filho, D. Brodzki, G. Djéga-Mariadassou, Formation of alkanes, alkylcycloalkanes and alkylbenzenes during the catalytic hydrocracking of vegetable oils, Fuel 72 (4) (1993) 543–549.
[384] A. Demirbaş, H. Kara, New options for conversion of vegetable oils to alternative fuels, Energy Sources A 28 (7) (2006) 619–626.
[385] R.W. Gosselink, S.A. Hollak, S.W. Chang, J. van Haveren, K.P. de Jong, J.H. Bitter, et al., Reaction pathways for the deoxygenation of vegetable oils and related model compounds, ChemSusChem 6 (9) (2013) 1576–1594.
[386] R.W. Gosselink, W. Xia, M. Muhler, K.P. de Jong, J.H. Bitter, Enhancing the activity of Pd on carbon nanofibers for deoxygenation of amphiphilic fatty acid molecules through support polarity, ACS Catal. 3 (10) (2013) 2397–2402.
[387] Y. Qin, P. Chen, J. Duan, J. Han, H. Lou, X. Zheng, et al., Carbon nanofibers supported molybdenum carbide catalysts for hydrodeoxygenation of vegetable oils, RSC Adv. 3 (38) (2013) 17485–17491.
[388] S.D. Kushch, N.S. Kujunko, B.P. Tarasov, Platinum nanoparticles on carbon nanomaterials with graphene structure as hydrogenation catalysts, Russ. J. Gen. Chem. 79 (4) (2009) 706–710.
[389] J.J. Delgado, R. Vieira, G. Rebmann, D.S. Su, N. Keller, M.J. Ledoux, et al., Supported carbon nanofibers for the fixed-bed synthesis of styrene, Carbon. 44 (4) (2006) 809–812.
[390] N. Xiao, Y. Zhou, Z. Ling, Z. Zhao, J. Qiu, Carbon foams made of in situ produced carbon nanocapsules and the use as a catalyst for oxidative dehydrogenation of ethylbenzene, Carbon. 60 (2013) 514–522.
[391] N. Maximova, Partialdehydrierung von Ethylbenzol zu Styrol an Kohlenstoffmaterialien (Ph.D. thesis), Technische Universität Berlin, Fakultät II-Mathematik und Naturwissenschaften, (2003).
[392] T.J. Zhao, W.Z. Sun, X.Y. Gu, M. Rønning, D. Chen, Y.C. Dai, et al., Rational design of the carbon nanofiber catalysts for oxidative dehydrogenation of ethylbenzene, Appl. Catal. A 323 (2007) 135–146.
[393] R.D. Weinstein, A.R. Ferens, R.J. Orange, P. Lemaire, Oxidative dehydrogenation of ethanol to acetaldehyde and ethyl acetate by graphite nanofibers, Carbon 49 (2) (2011) 701–707.
[394] Z.J. Sui, J.H. Zhou, Y.C. Dai, W.K. Yuan, Oxidative dehydrogenation of propane over catalysts based on carbon nanofibers, Catal. Today 106 (1) (2005) 90–94.
[395] E.S. Steigerwalt, G.A. Deluga, D.E. Cliffel, C.M. Lukehart, A Pt-Ru/graphitic carbon nanofiber nanocomposite exhibiting high relative performance as a direct-methanol fuel cell anode catalyst, J. Phys. Chem. B. 105 (34) (2001) 8097–8101.
[396] Yoshida, W., Nature Climate Change, May 2016.
[397] A. Kalra, S. Garde, G. Hummer, Osmotic water transport through carbon nanotube membranes, Proc. Natl. Acad. Sci. USA 100 (18) (2003) 10175–10180.

[398] G. Hummer, J.C. Rasaiah, J.P. Noworyta, Water conduction through the hydrophobic channel of a carbon nanotube, Nature. 414 (6860) (2001) 188–190.
[399] Q. Nan, P. Li, B. Cao, Fabrication of positively charged nanofiltration membrane via the layer-by-layer assembly of graphene oxide and polyethylenimine for desalination, Appl. Surf. Sci. 387 (2016) 521–528.
[400] Dow Chemical Company—Data sheet for NF-4040.
[401] P. Sun, R. Ma, W. Ma, J. Wu, K. Wang, T. Sasaki, et al., Highly selective charge-guided ion transport through a hybrid membrane consisting of anionic graphene oxide and cationic hydroxide nanosheet superlattice units, NPG Asia Mater. 8 (4) (2016) e259.
[402] Z. Jia, W. Shi, Tailoring permeation channels of graphene oxide membranes for precise ion separation, Carbon 101 (2016) 290–295.
[403] S. Park, K.S. Lee, G. Bozoklu, W. Cai, S.T. Nguyen, R.S. Ruoff, Graphene oxide papers modified by divalent ions—enhancing mechanical properties via chemical cross-linking, ACS Nano 2 (3) (2008) 572–578.
[404] Z. An, O.C. Compton, K.W. Putz, L.C. Brinson, S.T. Nguyen, Bio-inspired borate cross-linking in ultra-stiff graphene oxide thin films, Adv. Mater. 23 (33) (2011) 3842–3846.
[405] Y. Tian, Y. Cao, Y. Wang, W. Yang, J. Feng, Realizing ultrahigh modulus and high strength of macroscopic graphene oxide papers through crosslinking of mussel-inspired polymers, Adv. Mater. 25 (21) (2013) 2980–2983.
[406] Z. Jia, Y. Wang, Covalently crosslinked graphene oxide membranes by esterification reactions for ions separation, J. Mater. Chem. A 3 (8) (2015) 4405–4412.
[407] D. Li, X. Zhang, G.P. Simon, H. Wang, Forward osmosis desalination using polymer hydrogels as a draw agent: influence of draw agent, feed solution and membrane on process performance, Water Res. 47 (1) (2013) 209–215.
[408] A. Razmjou, G.P. Simon, H. Wang, Effect of particle size on the performance of forward osmosis desalination by stimuli-responsive polymer hydrogels as a draw agent, Chem. Eng. J. 215 (2013) 913–920.
[409] D. Li, X. Zhang, J. Yao, G.P. Simon, H. Wang, Stimuli-responsive polymer hydrogels as a new class of draw agent for forward osmosis desalination, Chem. Commun. 47 (6) (2011) 1710–1712.
[410] Y. Cai, W. Shen, S.L. Loo, W.B. Krantz, R. Wang, A.G. Fane, et al., Towards temperature driven forward osmosis desalination using Semi-IPN hydrogels as reversible draw agents, Water Res. 47 (11) (2013) 3773–3781.
[411] D. Li, X. Zhang, J. Yao, Y. Zeng, G.P. Simon, H. Wang, Composite polymer hydrogels as draw agents in forward osmosis and solar dewatering, Soft Matter 7 (21) (2011) 10048–10056.
[412] A. Razmjou, Q. Liu, G.P. Simon, H. Wang, Bifunctional polymer hydrogel layers as forward osmosis draw agents for continuous production of fresh water using solar energy, Environ. Sci. Technol. 47 (22) (2013) 13160–13166.
[413] Y. Zeng, L. Qiu, K. Wang, J. Yao, D. Li, G.P. Simon, et al., Significantly enhanced water flux in forward osmosis desalination with polymer-graphene composite hydrogels as a draw agent, RSC Advances 3 (3) (2013) 887–894.
[414] F. Banat, N. Jwaied, Exergy analysis of desalination by solar-powered membrane distillation units, Desalination 230 (1) (2008) 27–40.
[415] A.M. Alklaibi, N. Lior, Membrane-distillation desalination: status and potential, Desalination 171 (2) (2005) 111–131.
[416] S.T. Hsu, K.T. Cheng, J.S. Chiou, Seawater desalination by direct contact membrane distillation, Desalination 143 (3) (2002) 279–287.
[417] J.B. Gálvez, L. García-Rodríguez, I. Martín-Mateos, Seawater desalination by an innovative solar-powered membrane distillation system: the MEDESOL project, Desalination 246 (1) (2009) 567–576.

[418] H.E. Fath, S.M. Elsherbiny, A.A. Hassan, M. Rommel, M. Wieghaus, J. Koschikowski, et al., PV and thermally driven small-scale, stand-alone solar desalination systems with very low maintenance needs, Desalination 225 (1) (2008) 58–69.
[419] Z. Ding, L. Liu, M.S. El-Bourawi, R. Ma, Analysis of a solar-powered membrane distillation system, Desalination 172 (1) (2005) 27–40.
[420] J. Koschikowski, M. Wieghaus, M. Rommel, Solar thermal driven desalination plants based on membrane distillation, Water Sci. Technol. Water Supply 3 (5–6) (2003) 49–55.
[421] J. Walton, H. Lu, C. Turner, S. Solis, H. Hein, Solar and waste heat desalination by membrane distillation, in: Desalination and Water Purification Research and Development Program Report No. 81, 2004, 20.
[422] D. Qu, J. Wang, B. Fan, Z. Luan, D. Hou, Study on concentrating primary reverse osmosis retentate by direct contact membrane distillation, Desalination 247 (1) (2009) 540–550.
[423] L. Martinez-Diez, F.J. Florido-Diaz, Desalination of brines by membrane distillation, Desalination 137 (1) (2001) 267–273.
[424] M. Gryta, K. Karakulski, The application of membrane distillation for the concentration of oil-water emulsions, Desalination 121 (1) (1999) 23–29.
[425] M. Gryta, M. Tomaszewska, K. Karakulski, Wastewater treatment by membrane distillation, Desalination 198 (1) (2006) 67–73.
[426] V. Calabrò, E. Drioli, F. Matera, Membrane distillation in the textile wastewater treatment, Desalination 83 (1–3) (1991) 209–224.
[427] C. Boi, S. Bandini, G.C. Sarti, Pollutants removal from wastewaters through membrane distillation, Desalination 183 (1) (2005) 383–394.
[428] N. Wang, S. Ji, G. Zhang, J. Li, L. Wang, Self-assembly of graphene oxide and polyelectrolyte complex nanohybrid membranes for nanofiltration and pervaporation, Chem. Eng. J. 213 (2012) 318–329.
[429] N. Wang, S. Ji, J. Li, R. Zhang, G. Zhang, Poly (vinyl alcohol)–grapheme oxide nanohybrid "pore-filling" membrane for pervaporation of toluene/n-heptane mixtures, J. Memb. Sci. 455 (2014) 113–120.
[430] B. Liang, W. Zhan, G. Qi, S. Lin, Q. Nan, Y. Liu, et al., High performance graphene oxide/polyacrylonitrile composite pervaporation membranes for desalination applications, J. Mater. Chem. A 3 (9) (2015) 5140–5147.
[431] J. Zhao, Y. Zhu, F. Pan, G. He, C. Fang, K. Cao, et al., Fabricating graphene oxide-based ultrathin hybrid membrane for pervaporation dehydration via layer-by-layer self-assembly driven by multiple interactions, J. Memb. Sci. 487 (2015) 162–172.
[432] D.P. Suhas, A.V. Raghu, H.M. Jeong, T.M. Aminabhavi, Graphene-loaded sodium alginate nanocomposite membranes with enhanced isopropanol dehydration performance via a pervaporation technique, RSC Advances 3 (38) (2013) 17120–17130.
[433] A. Yamasaki, T. Shinbo, K. Mizoguchi, Pervaporation of benzene/cyclohexane and benzene/n-hexane mixtures through PVA membranes, J. Appl. Polym. Sci. 64 (6) (1997) 1061–1065.
[434] G. Zhao, J. Li, X. Ren, C. Chen, X. Wang, Few-layered graphene oxide nanosheets as superior sorbents for heavy metal ion pollution management, Environ. Sci. Technol. 45 (24) (2011) 10454–10462.
[435] V. Chandra, J. Park, Y. Chun, J.W. Lee, I.C. Hwang, K.S. Kim, Water-dispersible magnetite-reduced graphene oxide composites for arsenic removal, ACS Nano 4 (7) (2010) 3979–3986.
[436] S.T. Yang, Y. Chang, H. Wang, G. Liu, S. Chen, Y. Wang, et al., Folding/aggregation of graphene oxide and its application in Cu^{2+} removal, J. Colloid. Interface Sci. 351 (1) (2010) 122–127.

[437] Y.C. Lee, J.W. Yang, Self-assembled flower-like TiO_2 on exfoliated graphite oxide or heavy metal removal, J. Ind. Eng. Chem. 18 (3) (2012) 1178–1185.
[438] M. Liu, C. Chen, J. Hu, X. Wu, X. Wang, Synthesis of magnetite/graphene oxide composite and application for cobalt (II) removal, J. Phys. Chem. C 115 (51) (2011) 25234–25240.
[439] C.J. Madadrang, H.Y. Kim, G. Gao, N. Wang, J. Zhu, H. Feng, et al., Adsorption behavior of EDTA-graphene oxide for Pb (II) removal, ACS Appl. Mater. Interfaces 4 (3) (2012) 1186–1193.
[440] X. Mi, G. Huang, W. Xie, W. Wang, Y. Liu, J. Gao, Preparation of graphene oxide aerogel and its adsorption for Cu^{2+} ions, Carbon 50 (13) (2012) 4856–4864.
[441] J.H. Deng, X.R. Zhang, G.M. Zeng, J.L. Gong, Q.Y. Niu, J. Liang, Simultaneous removal of Cd (II) and ionic dyes from aqueous solution using magnetic graphene oxide nanocomposite as an adsorbent, Chem. Eng. J. 226 (2013) 189–200.
[442] Y. Yuan, G. Zhang, Y. Li, G. Zhang, F. Zhang, X. Fan, Poly (amidoamine) modified graphene oxide as an efficient adsorbent for heavy metal ions, Polym. Chem. 4 (6) (2013) 2164–2167.
[443] L. Fan, C. Luo, M. Sun, H. Qiu, Synthesis of graphene oxide decorated with magnetic cyclodextrin for fast chromium removal, J. Mater. Chem. 22 (47) (2012) 24577–24583.
[444] G. Gollavelli, C.C. Chang, Y.C. Ling, Facile synthesis of smart magnetic graphene for safe drinking water: heavy metal removal and disinfection control, ACS Sustainable Chem. Eng. 1 (5) (2013) 462–472.
[445] W. Li, S. Gao, L. Wu, S. Qiu, Y. Guo, X. Geng, et al., High-density three-dimension graphene macroscopic objects for high-capacity removal of heavy metal ions, Sci. Rep. 3 (2013). Article number: 2125.
[446] W. Zhang, X. Shi, Y. Zhang, W. Gu, B. Li, Y. Xian, Synthesis of water-soluble magnetic graphene nanocomposites for recyclable removal of heavy metal ions, J. Mater. Chem. A 1 (5) (2013) 1745–1753.
[447] N. Zhang, H. Qiu, Y. Si, W. Wang, J. Gao, Fabrication of highly porous biodegradable monoliths strengthened by graphene oxide and their adsorption of metal ions, Carbon 49 (3) (2011) 827–837.
[448] G. Zhao, J. Li, X. Ren, C. Chen, X. Wang, Few-layered graphene oxide nanosheets as superior sorbents for heavy metal ion pollution management, Environ. Sci. Technol. 45 (24) (2011) 10454–10462.
[449] I.E.M. Carpio, J.D. Mangadlao, H.N. Nguyen, R.C. Advincula, D.F. Rodrigues, Graphene oxide functionalized with ethylenediamine triacetic acid for heavy metal adsorption and anti-microbial applications, Carbon 77 (2014) 289–301.
[450] L. Fan, C. Luo, M. Sun, X. Li, H. Qiu, Highly selective adsorption of lead ions by water-dispersible magnetic chitosan/graphene oxide composites, Colloids Surf. B Biointerfaces 103 (2013) 523–529.
[451] Z. Dong, D. Wang, X. Liu, X. Pei, L. Chen, J. Jin, Bio-inspired surface-functionalization of graphene oxide for the adsorption of organic dyes and heavy metal ions with a superhigh capacity, J. Mater. Chem. A 2 (14) (2014) 5034–5040.
[452] R. Sitko, E. Turek, B. Zawisza, E. Malicka, E. Talik, J. Heimann, et al., Adsorption of divalent metal ions from aqueous solutions using grapheme oxide, Dalton Trans. 42 (16) (2013) 5682–5689.
[453] Y.L.F. Musico, C.M. Santos, M.L.P. Dalida, D.F. Rodrigues, Improved removal of lead (II) from water using a polymer-based graphene oxide nanocomposite, J. Mater. Chem. A 1 (11) (2013) 3789–3796.
[454] W. Wu, Y. Yang, H. Zhou, T. Ye, Z. Huang, R. Liu, et al., Highly efficient removal of Cu (II) from aqueous solution by using graphene oxide, Water Air Soil Pollut. 224 (1) (2013) 1–8.

[455] S. Luo, X. Xu, G. Zhou, C. Liu, Y. Tang, Y. Liu, Amino siloxane oligomer-linked graphene oxide as an efficient adsorbent for removal of Pb (II) from wastewater, J. Hazard. Mater. 274 (2014) 145–155.
[456] L.P. Lingamdinne, J.R. Koduru, H. Roh, Y.L. Choi, Y.Y. Chang, J.K. Yang, Adsorption removal of Co (II) from waste-water using graphene oxide, Hydrometallurgy. 165 (2015) 90–96.
[457] H. Qin, T. Gong, Y. Cho, C. Lee, T. Kim, A conductive copolymer of graphene oxide/poly (1-(3-aminopropyl) pyrrole) and the adsorption of metal ions, Polym. Chem. 5 (15) (2014) 4466–4473.
[458] M. Tan, X. Liu, W. Li, H. Li, Enhancing sorption capacities for copper (II) and lead (II) under weakly acidic conditions by L-tryptophan-functionalized graphene oxide, J. Chem. Eng. Data 60 (5) (2015) 1469–1475.
[459] G. Zhao, X. Ren, X. Gao, X. Tan, J. Li, C. Chen, et al., Removal of Pb (II) ions from aqueous solutions on few-layered graphene oxide nanosheets, Dalton Trans. 40 (41) (2011) 10945–10952.
[460] V.P. Chauke, A. Maity, A. Chetty, High-performance towards removal of toxic hexavalent chromium from aqueous solution using graphene oxide-alpha cyclodextrin-polypyrrole nanocomposites, J. Mol. Liquids 211 (2015) 71–77.
[461] Y. Zhang, Y. Liu, X. Wang, Z. Sun, J. Ma, T. Wu, et al., Porous graphene oxide/carboxymethyl cellulose monoliths, with high metal ion adsorption, Carbohydr. Polym. 101 (2014) 392–400.
[462] C. Jiao, J. Xiong, J. Tao, S. Xu, D. Zhang, H. Lin, et al., Sodium alginate/graphene oxide aerogel with enhanced strength–toughness and its heavy metal adsorption study, Int. J. Biol. Macromol. 83 (2016) 133–141.
[463] Z. Dong, F. Zhang, D. Wang, X. Liu, J. Jin, Polydopamine-mediated surface-functionalization of graphene oxide for heavy metal ions removal, J. Solid State Chem. 224 (2015) 88–93.
[464] S. Wan, F. He, J. Wu, W. Wan, Y. Gu, B. Gao, Rapid and highly selective removal of lead from water using graphene oxide-hydrated manganese oxide nanocomposites, J. Hazard. Mater. 314 (2016) 32–40.
[465] J. Li, S. Zhang, C. Chen, G. Zhao, X. Yang, J. Li, et al., Removal of Cu (II) and fulvic acid by graphene oxide nanosheets decorated with Fe_3O_4 nanoparticles, ACS Appl. Mater. Interfaces 4 (9) (2012) 4991–5000.
[466] Chelation Resin Booklet, Purolite Company, PA.
[467] T. Pellenbarg, N. Dementev, R. Jean-Gilles, C. Bessel, E. Borguet, N. Dollahon, et al., Detecting and quantifying oxygen functional groups on graphite nanofibers by fluorescence labeling of surface species, Carbon 48 (15) (2010) 4256–4267.
[468] K. Zhang, V. Dwivedi, C. Chi, J. Wu, Graphene oxide/ferric hydroxide composites for efficient arsenate removal from drinking water, J. Hazard. Mater. 182 (1) (2010) 162–168.
[469] A.K. Mishra, S. Ramaprabhu, Functionalized graphene sheets for arsenic removal and desalination of sea water, Desalination 282 (2011) 39–45.
[470] G. Sheng, Y. Li, X. Yang, X. Ren, S. Yang, J. Hu, et al., Efficient removal of arsenate by versatile magnetic graphene oxide composites, RSC Advances 2 (32) (2012) 12400–12407.
[471] T. Wen, X. Wu, X. Tan, X. Wang, A. Xu, One-pot synthesis of water-swellable Mg–Al layered double hydroxides and graphene oxide nanocomposites for efficient removal of As (V) from aqueous solutions, ACS Appl. Mater. Interfaces 5 (8) (2013) 3304–3311.
[472] L. Li, G. Zhou, Z. Weng, X.Y. Shan, F. Li, H.M. Cheng, Monolithic Fe_2O_3 3/graphene hybrid for highly efficient lithium storage and arsenic removal, Carbon 67 (2014) 500–507.

[473] X. Luo, C. Wang, L. Wang, F. Deng, S. Luo, X. Tu, et al., Nanocomposites of graphene oxide-hydrated zirconium oxide for simultaneous removal of As (III) and As (V) from water, Chem. Eng. J. 220 (2013) 98−106.

[474] J. Zhu, R. Sadu, S. Wei, D.H. Chen, N. Haldolaarachchige, Z. Luo, et al., Magnetic graphene nanoplatelet composites toward arsenic removal, ECS J. Solid State Sci. Technol. 1 (1) (2012) M1−M5.

[475] X.L. Wu, L. Wang, C.L. Chen, A.W. Xu, X.K. Wang, Water-dispersible magnetite-graphene-LDH composites for efficient arsenate removal, J. Mater. Chem. 21 (43) (2011) 17353−17359.

[476] S. Kumar, R.R. Nair, P.B. Pillai, S.N. Gupta, M.A.R. Iyengar, A.K. Sood, Graphene oxide−$MnFe_2O_4$ magnetic nanohybrids for efficient removal of lead and arsenic from water, ACS Appl. Mater. Interfaces 6 (20) (2014) 17426−17436.

[477] C. Wang, H. Luo, Z. Zhang, Y. Wu, J. Zhang, S. Chen, Removal of As (III) and As (V) from aqueous solutions using nanoscale zero valent iron-reduced graphite oxide modified composites, J. Hazard. Mater. 268 (2014) 124−131.

[478] M.L. Chen, Y. Sun, C.B. Huo, C. Liu, J.H. Wang, Akaganeite decorated graphene oxide composite for arsenic adsorption/removal and its proconcentration at ultra-trace level, Chemosphere 130 (2015) 52−58.

[479] O.S. Thirunavukkarasu, T. Viraraghavan, K.S. Subramanian, Arsenic removal from drinking water using iron oxide-coated sand, Water Air Soil Pollut. 142 (1−4) (2003) 95−111.

[480] R.R. Devi, I.M. Umlong, B. Das, K. Borah, A.J. Thakur, P.K. Raul, et al., Removal of iron and arsenic (III) from drinking water using iron oxide-coated sand and limestone, Appl. Water Sci. 4 (2) (2014) 175−182.

[481] G. Crini, H.N. Peindy, Adsorption of CI Basic Blue 9 on cyclodextrin-based material containing carboxylic groups, Dyes Pigm. 70 (3) (2006) 204−211.

[482] N.A. Travlou, G.Z. Kyzas, N.K. Lazaridis, E.A. Deliyanni, Graphite oxide/chitosan composite for reactive dye removal, Chem. Eng. J. 217 (2013) 256−265.

[483] J.A. González, M.E. Villanueva, L.L. Piehl, G.J. Copello, Development of a chitin/graphene oxide hybrid composite for the removal of pollutant dyes: adsorption and desorption study, Chem. Eng. J. 280 (2015) 41−48.

[484] T. Jiao, Y. Liua, Y. Wua, Q. Zhanga, X. Yana, F. Gaoa, A.J.P. Bauera, et al., Facile and scalable preparation of graphene oxide-based magnetic hybrids for fast and highly efficient removal of organic dyes, Sci. Rep. 5 (2015). Article number: 12451.

[485] S. Shahabuddin, N.M. Sarih, M. Afzal Kamboh, H. Rashidi Nodeh, S. Mohamad, Synthesis of polyaniline-coated graphene oxide@ $SrTiO_3$ nanocube nanocomposites for enhanced removal of carcinogenic dyes from aqueous solution, Polymers 8 (9) (2016) 305.

[486] L. Ji, W. Chen, Z. Xu, S. Zheng, D. Zhu, Graphene nanosheets and graphite oxide as promising adsorbents for removal of organic contaminants from aqueous solution, J. Environ. Qual. 42 (1) (2013) 191−198.

[487] C. Park, E.S. Engel, A. Crowe, T.R. Gilbert, N.M. Rodriguez, Use of carbon nanofibers in the removal of organic solvents from water, Langmuir. 16 (21) (2000) 8050−8056.

[488] N. Pugazhenthiran, S. Sen Gupta, A. Prabhath, M. Manikandan, J.R. Swathy, V.K. Raman, et al., Cellulose derived graphenic fibers for capacitive desalination of brackish water, ACS Appl. Mater. Interfaces 7 (36) (2015) 20156−20163.

[489] H. Li, L. Pan, C. Nie, Y. Liu, Z. Sun, Reduced graphene oxide and activated carbon composites for capacitive deionization, J. Mater. Chem. 22 (31) (2012) 15556−15561.

[490] Y. Gendel, N.O. Levi, O. Lahav, H_2S (g) removal using a modified, low-pH liquid redox sulfur recovery (LRSR) process with electrochemical regeneration of the Fe catalyst couple, Environ. Sci. Technol. 43 (21) (2009) 8315–8319.
[491] G.J. Nagl, Liquid redox enhances Claus process, Sulfur May-June (274) (2001).
[492] B. Huang, B.B. Chen, R. Chen, Predicting H_2S oxidative dehydrogenation over graphene oxides from first principles, Chin. J. Chem. Phys. 28 (2) (2015) 143–149.
[493] J.C. Aguila, H.H. Cocoletzi, G.H. Cocoletzi, A theoretical analysis of the role of defects in the adsorption of hydrogen sulfide on graphene, AIP Advances 3 (3) (2013) 032118.
[494] S. Gadipelli, Z.X. Guo, Graphene-based materials: synthesis and gas sorption, storage and separation, Prog. Mater. Sci. 69 (2015) 1–60.
[495] J.R. Li, R.J. Kuppler, H.C. Zhou, Selective gas adsorption and separation in metal–organic frameworks, Chem. Soc. Rev. 38 (5) (2009) 1477–1504.
[496] R.T. Yang, Adsorbents: Fundamentals and Applications, John Wiley & Sons, Hoboken, NJ, 2003.
[497] S.O. Aloba, Carbon dioxide capture by functionalized graphene oxide adsorbent (Doctoral dissertation), The University of Mississippi, 2015.
[498] D. Iruretagoyena, M.S. Shaffer, D. Chadwick, Adsorption of carbon dioxide on graphene oxide supported layered double oxides, Adsorption 20 (2–3) (2014) 321–330.
[499] A.K. Mishra, S. Ramaprabhu, Carbon dioxide adsorption in grapheme sheets, AIP Advances 1 (3) (2011) 032152.
[500] R.R. Judkins, T.D. Burchell, Gas separation device based on electrical swing adsorption, US Patent 5,972, 077, 1999.
[501] S.U.S. Choi, Z.G. Zhang, W. Yu, F.E. Lockwood, E.A. Grulke, Anomalous thermal conductivity enhancement in nanotube suspensions, Appl. Phys. Lett. 79 (14) (2001) 2252–2254.
[502] P. Estelle, S. Halelfadl, T. Mare, Thermophysical properties and heat transfer performance of carbon nanotubes water-based nanofluids, J. Therm. Anal. Calorim., Springer-Verlag; Springer (Kluwer Academic Publishers) (2016) <http://dx.doi.org/10.1007/s10973-016-5833-8>, <hal-01368812>.
[503] M. Piratheepan, T.N. Anderson, An experimental investigation of turbulent forced convection heat transfer by a multi-walled carbon-nanotube nanofluid, Int. Commun. Heat Mass Transfer 57 (2014) 286–290.
[504] E. Hosseinipour, S.Z. Heris, M. Shanbedi, Experimental investigation of pressure drop and heat transfer performance of amino acid-functionalized MWCNT in the circular tube, J. Therm. Anal. Calorim. 124 (1) (2016) 205–214.
[505] Y. Ding, H. Alias, D. Wen, R.A. Williams, Heat transfer of aqueous suspensions of carbon nanotubes (CNT nanofluids), Int. J. Heat Mass Transfer 49 (1) (2006) 240–250.
[506] W. Yu, H. Xie, W. Chen, Experimental investigation on thermal conductivity of nanofluids containing graphene oxide nanosheets, J. Appl. Phys. 107 (9) (2010) 094317.
[507] A. Nasiri, M. Shariaty-Niasar, A. Rashidi, A. Amrollahi, R. Khodafarin, Effect of dispersion method on thermal conductivity and stability of nanofluid, Exp. Therm. Fluid Sci. 35 (4) (2011) 717–723.
[508] W. Zhang, M. Zhou, H. Zhu, Y. Tian, K. Wang, J. Wei, et al., Tribological properties of oleic acid-modified graphene as lubricant oil additives, J. Phys. D Appl. Phys. 44 (20) (2011) 205303.
[509] V. Eswaraiah, V. Sankaranarayanan, S. Ramaprabhu, Graphene-based engine oil nanofluids for tribological applications, ACS Appl. Mater. Interfaces 3 (11) (2011) 4221–4227.

[510] H.D. Huang, J.P. Tu, L.P. Gan, C.Z. Li, An investigation on tribological properties of graphite nanosheets as oil additive, Wear 261 (2) (2006) 140–144.
[511] A. Senatore, V. D'Agostino, V. Petrone, P. Ciambelli, M. Sarno, Graphene oxide nanosheets as effective friction modifier for oil lubricant: materials, methods, and tribological results, ISRN Tribol. 2013 (2013)9 pp.
[512] K.H. Park, B. Ewald, P.Y. Kwon, Effect of nano-enhanced lubricant in minimum quantity lubrication balling milling, J. Tribol. 133 (3) (2011) 031803.
[513] P.S. Sreejith, B.K.A. Ngoi, Dry machining: machining of the future, J. Mater. Process. Technol. 101 (1) (2000) 287–291.
[514] F. Klocke, G. Eisenblätter, Dry cutting, CIRP Annals-Manuf. Technol. 46 (2) (1997) 519–526.
[515] D.V. Kosynkin, G. Ceriotti, K.C. Wilson, J.R. Lomeda, J.T. Scorsone, A.D. Patel, et al., Graphene oxide as a high-performance fluid-loss-control additive in water-based drilling fluids, ACS Appl. Mater. Interfaces 4 (1) (2011) 222–227.
[516] Functionalized graphene oxide plays part in next generation oil-well drilling fluids. <http://www.eurekalert.org/pub_releases/2011-12/ru-fgo120811.php>.
[517] A.R. Ismail, W. Sulaiman, W. Rosli, M.Z. Jaafar, I. Ismail, E. Sabu Hera, Nanoparticles performance as fluid loss additives in water based drilling fluids, Mater. Sci. Forum 864 (2016) 189–193.
[518] N.M. Taha, S. Lee, Nano graphene application improving drilling fluids performance, in: International Petroleum Technology Conference, December 2015.
[519] J.B. Fontenot, Lost circulation drilling fluids comprising elastomeric rubber particles and a method for decreasing whole mud loss using such composition, US Patent 2015/0008044 A1, 2015.
[520] Y.H. Chai, S. Yusup, V.S. Chok, A review on nanoparticle addition in base fluid for improvement of biodegradable ester-based drilling fluid properties, Chem. Eng. 45 (2015) 1447–1452.
[521] L. Jassim, R. Yunus, U. Rashid, S.A. Rashid, M.A. Salleh, S. Irawan, et al., Synthesis and optimization of 2-ethylhexyl ester as base oil for drilling fluid formulation, Chem. Eng. Commun. 203 (4) (2016) 463–470.

INDEX

Note: Page numbers followed by "*f*" and "*t*" refer to figures and tables, respectively.

F

Q

R

S

T

Printed in the United States
By Bookmasters